Microsoft Fabric Handbook

Guide for Overview, Policy, Governance, AI Adoption and Solution Deployment Leveraging Microsoft Fabric

By

Claude Louis-Charles, PhD

Todd Pliner

Managers Guide to Microsoft Fabric

The moral rights of the author has been asserted

Copyrights © 2024

All rights reserved-Claude J Louis-Charles, PhD

No part of this publication may be reproduced, stored in a retrieval system or transmitted in any form or by any means, without the prior permission in writing of the publisher, nor be otherwise circulated in any form of binding or cover other than that in which it is published and without a similar condition including this condition being imposed on the subsequent purchaser.

Cybersoft Publishing LLC
Fort Washington Md 20744

https://www.drclaude.net

Illustrations by Kyasia Morgan

First Edition October 2024

Acknowledgments

We would like to express my gratitude to our peers that unknowingly have been serving as our sound board. We truly appreciated their invaluable patience and feedback for our endless series of questions and ideas.

I (Claude) am endlessly thankful to my loving wife, Tracy, who has supported me for over 30 years, and my children, JC and Jordan, who have been cheering me on no matter how many books I choose to write. Their unwavering belief in me has kept my spirits and motivation high throughout this lengthy process.

I (Todd) am also very thankful to my loving wife.

TABLE OF CONTENTS

Chapter 1
Introduction

Overview of the Books Purpose and Content

This book aims to provide managers and engineers with an overview of Microsoft Fabric, a comprehensive offering that includes numerous components. Microsoft Fabric is a unified data solution platform based on a Software as a Service (SaaS) cloud approach, integrating OneLake with various data storage capabilities, development tools, and analytics tools for organizations seeking a central solution. With the proliferation of artificial intelligence (AI) and the increased adoption of data science by corporations, data has become an essential asset for digital transformation and a critical tool for competitiveness. However, the previous duplication of data science tools often led to unnecessary complexities. Many firms had data science experts using various tools to gather, store, assess, display, and engineer data, resulting in disjointed and disorganized DataLakes that complicated their construction, integration, management, and use. To address these issues, Microsoft developed Microsoft Fabric to provide integrated and easy-to-adopt tools.

The modern Microsoft Fabric package constitutes an inclusive platform for data science through SaaS, offering data transition, processing, breakdown, conversion, live event routing, and report formulation capabilities.

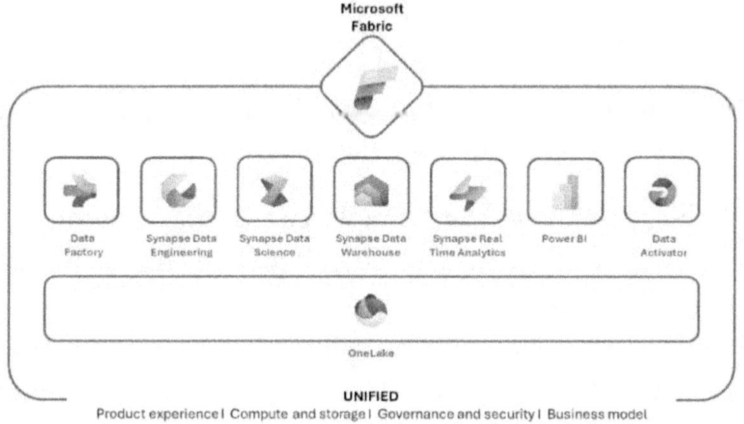

Figure 1 Fabric Overview

Microsoft Fabric is a centralized platform offering various components, from Power BI to real-time analytics tools, revolutionizing the data science field through AI integration. Fabric is also a comprehensive analytics platform that allows data experts and business users to collaborate on data projects. It offers a suite of connected services that enable you to collect, store, transform, and analyze data in one environment.

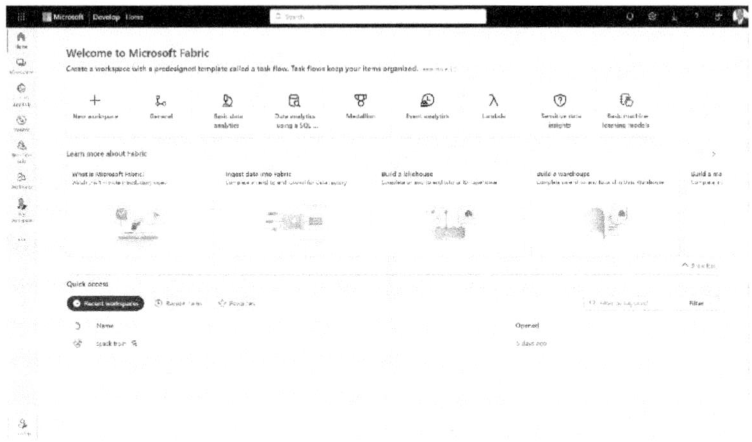

Figure 2 Main Fabric Screen

Fabric is user-friendly and simplifies analytics by providing everything you need in one place, eliminating the need to use multiple vendors. Microsoft Fabric combines different elements into a unified stack, allowing you to consolidate data storage with OneLake instead of using various databases or data warehouses. AI features are seamlessly incorporated within Fabric, eliminating the need for manual integration. With Fabric, you can effortlessly transform your raw data into useful insights for business users. Microsoft Fabric offers solutions for data practitioners of varying levels of expertise and connects with tools businesses require to make informed decisions.

The Evolutionary Nature of Microsoft Fabric

Microsoft Fabric is a SaaS platform that integrates both new and existing components from Power BI, Azure Synapse Analytics, Azure Data Factory, and other services into a cohesive environment. These components are then adapted into personalized user experiences.

Microsoft Fabric Architecture

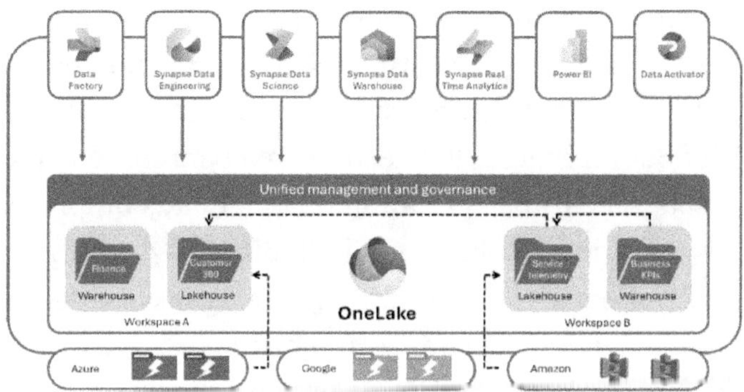

Figure 3 Fabric Architecture

A simple illustration of the Microsoft Fabric architecture comprises three basic layers: the bottom layer multi-cloud platform, the middle layer OneLake data storage center, and the top layer of workloads, as depicted in Figure 1 above. The bottom layer multi-cloud platform currently includes Amazon S3, Azure, and Google Cloud, with other cloud-based systems continuously being integrated. The multi-cloud scenario benefits enterprises relying on more than one cloud solution provider. Users can connect to these cloud solutions, pull data through available "shortcuts," and utilize them in their workflows. These "shortcuts" link the cloud solution to the middle layer, OneLake, a centralized data store. This storage layer is founded on the "data lakehouse" paradigm, integrating the best features of data lakes and warehouses. It serves as the main repository for the Fabric, storing all data in the 'data lake format' rather than the relational format. The data storage platform is open source, allowing users to incorporate products that can read from it. Additionally, it plays a crucial role in locating, exploring, and utilizing the multiple data assets within Microsoft Fabric. One of the key features of OneLake is that users can create shortcuts that direct them to other data stations, such as AWS S3, eliminating the need to create multiple copies of their assets. The top layer sitting on OneLake comprises the primary Fabric workloads, including Azure Synapse Analytics, Power BI, Azure Machine Learning, Data Activator, Synapse Data Warehousing, Synapse Data Engineering, and Synapse Real-Time Analytics.

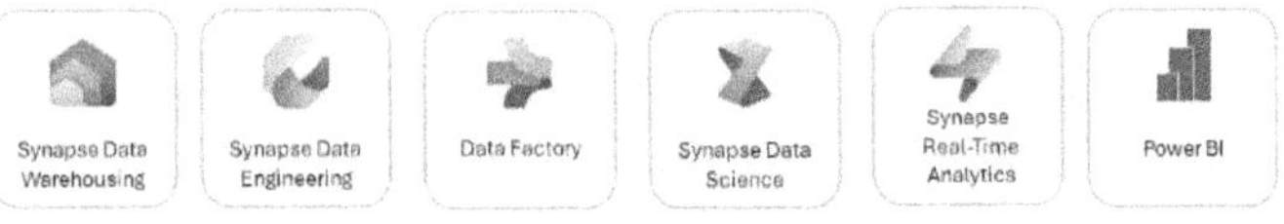

Figure 4 Workload List

1. **Synapse Data Warehouse**: Synapse Data Warehouse provides high-end SQL performance and scalability by decoupling computing from storage, allowing each to scale independently. It introduces a lake-centric data warehouse built on a business-grade distributed processing platform, delivering enterprise-leading performance at scale while eliminating the need for configuration and management. Through an easy-to-adopt SaaS experience, seamlessly integrated with Power BI for smooth evaluation and reporting, the warehouse bridges the gap between data lakes and warehouses, significantly simplifying a firm's analytics estate.

The associated workloads benefit from the extensive capabilities of the SQL engine through an open data structure, enabling clients to focus on preparing, analyzing, and reporting data using a single copy stored in Microsoft OneLake.

2. **Synapse Data Engineering**: Synapse Data Engineering provides a Spark service with exceptional authoring experience and data transformation capabilities. It allows individuals to create, organize, and optimize tools for gathering, storing, assessing, and analyzing large amounts of data. The Spark platform's integration with Data Factory enables businesses to schedule and execute notebooks and Spark tasks.

3. **Data Factory**: Data Factory offers a modern data integration platform for ingesting, preparing, and transforming data from a wide array of sources. It adopts the simplicity of Power Query and enables businesses to use more than two hundred native connectors to connect with data sources within enterprises and in the cloud.

4. **Synapse Data Science**: Synapse Data Science allows enterprises to construct, distribute, and activate machine learning (ML) techniques from Fabric. It connects with Azure ML to offer intrinsic experiment tracking and a model registry. Data scientists can use it to enrich enterprise data with forecasts, allowing business experts to include predictions in their BI reports. This approach

facilitates an organizational shift from descriptive to integrative insights.

5. **Synapse Real-Time Analytics (SRTA):** Synapse Real-Time Analytics (SRTA) is a structured big data analytics system optimized for streaming and time-series data, reducing data complexity and easing integration. The platform allows employees to gain insights and visualize data in motion, simplifying user experience while maintaining powerful analytical capabilities.

6. **Power BI**: Microsoft Fabric maximizes the capabilities of Power BI, a versatile business analytics tool that allows businesses to connect to their data sources, visualize, and share information. The workload supports interactive dashboards and numerous inbuilt data connectors, providing an integrated experience that enables enterprises to quickly and spontaneously access data in Fabric for enhanced decision-making. The workload's basic components include Power BI Desktop, SaaS Power BI Service, and mobile apps for iOS and Android products.

7. **Data Activator**: Data Activator is a zero-code experience in Fabric that enables users to pinpoint activities, such as email alerts, to initiate when it detects specific patterns in the continuously changing data. It assesses data in Power BI documents and event streams, taking recommended actions when certain thresholds are met or specific patterns are recognized.

In summary, Fabric consolidates different workloads for Data Engineering, Data Factory, Data Science, Data Warehouse, Real-Time Analytics, Data Activator, Industry Solutions, and Power BI. These workloads are tailored to different user roles, such as data engineers, scientists, or warehousing experts, each designed for a specific job. The entire Fabric stack is designed for AI integration, accelerating the data journey. These workloads work seamlessly together, offering several benefits:

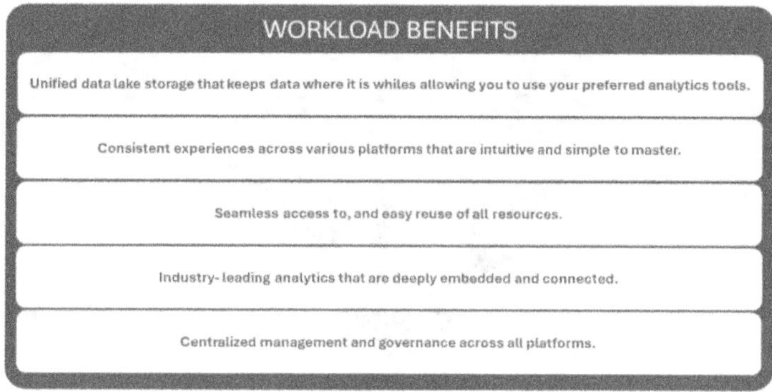

Figure 5 Workload Benefits

Fabric connects data and services smoothly, making them easy to manage, govern, and discover. It secures items, data, and row-level access, allowing centralized setup of core enterprise features. All underlying services follow the same permissions, and data sensitivity labels automatically apply to all suite items. Purview, part of Fabric, enables governance, allowing creators to focus on delivering their best work without worrying about the integration, management, or understanding of the supporting infrastructure.

The "Too Much Data" Problem

Many organizations face challenges tied to multiple sources of the same data. The concept of Shadow IT also plays a role in data duplication and security, which refers to software, hardware, or IT resources used on an enterprise network without the IT department's approval, knowledge, or oversight.

Figure 6 Too Much Data from Multiple Sources in Multiple Formats

AI Integration in Microsoft Fabric

Two options for utilizing Azure services provided by Fabric include pre-built AI prototypes and Bring your own key (BYOK). Fabric seamlessly interacts with Azure AI products, enabling users to enhance their data by utilizing access to native Azure AI models and prototypes. Organizations are encouraged to follow this standard since Fabric authentication persists through Azure services to access available AI products. According to Microsoft Build (2024), this offering is available for public preview with a few AI products. The second option, BYOK, allows enterprises to provide their AI services on Fabric and introduce their own key for use. If the needed AI services are not offered in the built-in AI prototypes, businesses can still utilize BYOK. Currently, GPT-3.5 available through Fabric Azure OpenAI can help businesses comprehend and produce natural language code. The GPT-3.5 Turbo is its most versatile version, optimized for chat and traditional task accomplishments.

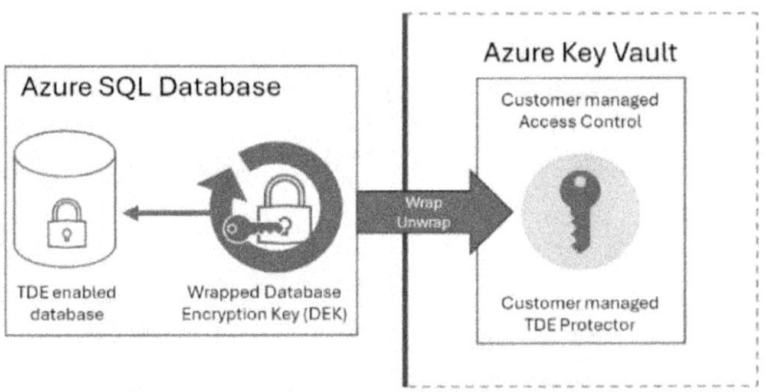

Figure 7 Bring Your Own Key (BYOK)

Impact of Microsoft Fabric on Generative AI and AI

The adoption of Microsoft Fabric significantly impacts Generative AI (GenAI) and AI utilization in general by assisting in data standardization, linking with AI tools and services, and easing data access. Businesses will benefit immensely from AI integration into Microsoft Fabric, simplifying data access without needing to move or copy it.

According to PWC (2024), most firms use data sets with varying standards in diverse systems, complicating AI tool access. However, Microsoft Fabric's universal platform will allow its data lake to integrate shortcuts pointing to other storage positions. Once connections are established, GenAI can easily access the data. Although the data stored in various locations may be in different formats, Fabric can help in standardization, troubleshooting, and organization, enabling GenAI applications to work from compatible and authoritative data sets. Additionally, as varying GenAI apps gain insights from Fabric data and introduce new data, Fabric can facilitate synchronization.

Sensors and Their Impact on Big Data

The world is in the information age, where sensors dominate life, industrial, and business activities.

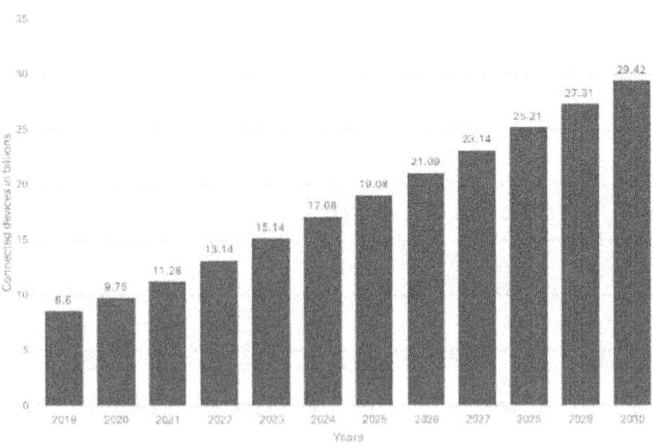

Figure 8 Connected Devices Growth

- Major industry verticals with over 100 million connected IoT devices include electricity, gas, steam & A/C, water supply & waste management, retail & wholesale, transportation & storage, and government.

- The consumer segment accounted for around 60% of all IoT connected devices in 2020.

- The most important use case for IoT devices in the consumer segment is consumer internet & media devices (such as smartphones), projected to grow to over 17 billion by 2030.

- Other use cases with more than one billion IoT devices by 2030 include connected (autonomous) vehicles, IT infrastructure, asset tracking & monitoring, and smart grids.

Data Collection and Utilization

Every device around you now collects vast quantities of data, from the watch on your arm to the multitude of data capture devices throughout your home. These sensors gather large amounts of data that can be used for various purposes at different levels. For example, businesses can use big data to facilitate relevant digital changes and achieve a competitive edge. Big data-driven analytics is gaining more prominence with the fast development and widespread adoption of artificial intelligence (AI) technologies. AI systems like language models and generative AI are crucial for businesses, as they help transform how workers use their time and how much data can be revealed.

Businesses need consistent, high-quality data and inferences from seamless analytics systems to get the best results from big data. However, many businesses rely on analytics systems that involve several separate and distinct services, making them more complex and expensive. Microsoft Fabric solves these issues by providing a complete and integrated analytics platform.

Azure Service Fabric

Microsoft's Azure Service Fabric is a platform for distributed systems that enable a microservice architecture. It simplifies packaging, deploying, and managing microservices and containers that are reliable and scalable (Sahay, 2020).

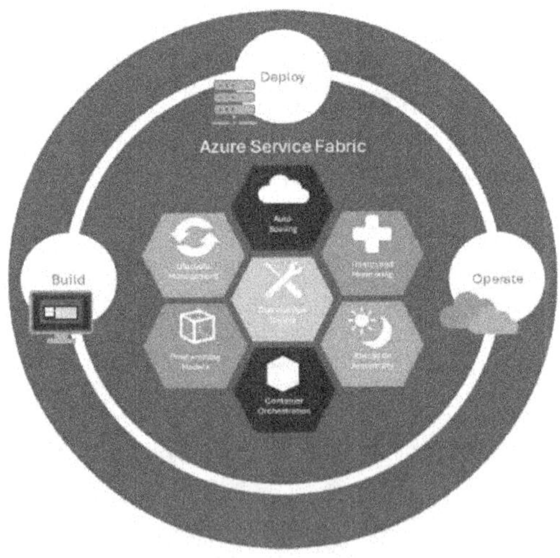

Figure 9 Service Fabric

The analytics platform also addresses the major challenges of building and running cloud-native applications.

Adoption by Leading Businesses

13

Various nationally and globally recognized businesses have adopted Microsoft Fabric to optimize analytics. For example, Ferguson, a leading North American waterworks, HVAC, and plumbing supplier, uses Fabric to consolidate its analytics (Ulagaratchagan, 2023). The objective is to improve efficiency and reduce delivery time. T-Mobile, one of the largest providers of wireless communication services, leverages Fabric as part of its transition to optimal data-driven decision-making (Ulagaratchagan, 2023). AON, which offers diverse services to its global customers, uses Fabric to consolidate its technology stack and add more value for its clients (Ulagaratchagan, 2023). These use cases indicate the increasing preference for a unified analytics platform.

Implications of Microsoft Fabric

Data Unification

Microsoft Fabric transforms the data science field from a disjointed sector, where data components and services are offered separately, to a unified platform where the tools are interacting in one environment. Fabric's revolutionary nature integrates both newer and existing tools, such as Data Factory and Data Activator. Fabric's integration with AI speeds data movement, allowing clients to access a wide range of intensively linked analytics tools. Other benefits of this unified system include quick access to and reuse of all assets, and central data lake storage that preserves data in its original location while utilizing the customer's selected analytics tools. Thus, Microsoft Fabric seamlessly links data and services, enabling unified organization, governance, administration, and detection. The platform guarantees security for assets, data, and low-level access by allowing central configuration of core enterprise capabilities.

Systems Cost Reduction

The unification of these systems under Fabric results in overall cost reductions, in contrast to traditional data analytics systems which integrate products from disparate vendors. This traditional approach leads to compute capacity being distributed across various systems to support data integration, transformations, and business intelligence reporting. However, when one system becomes idle, its capacity cannot be utilized by another system, resulting in considerable waste. With Microsoft Fabric, businesses can easily buy and manage resources through a single pool of compute resources that drive all Fabric workloads.

Liberation of Business Users

Fabric is also a powerful tool that empowers each enterprise user by allowing staff to make better decisions based on easily accessible data. Microsoft has fostered a data culture in firms by integrating Fabric with Microsoft 365 apps commonly used in businesses. Workloads such as Power BI have already been introduced into Microsoft 365. So customers using Microsoft 365 apps like PowerPoint and Excel can seamlessly discover data from OneLake, enabling them to derive more value. Microsoft 365 apps can seamlessly transform into hubs for discovering and applying insights. For example, individuals can use Excel to obtain and analyze data in OneLake and produce Power BI reports.

Empowerment Tool for Data Scientists

Azure OpenAI capabilities are integrated into every layer of Microsoft Fabric to ensure data professionals in organizations can accomplish more. Copilot within Fabric allows data scientists to use natural language to develop data flows, formulate SQL statements, construct reports, and create ML models. It also enables users to develop their own conversational language experiences, integrating Azure Open AI workflows with their data for publishing as plug-ins. The goal of Fabric's AI services is to empower businesses, even those without specialized AI or data science skills, by assisting programmers in creating applications with visual, auditory, speaking, understanding, and reasoning capabilities.

Data to Knowledge Process

Fabric, through its simplification and acceleration of data use, enables business users to quickly transition from raw data to intelligent decision tools like Power BI.

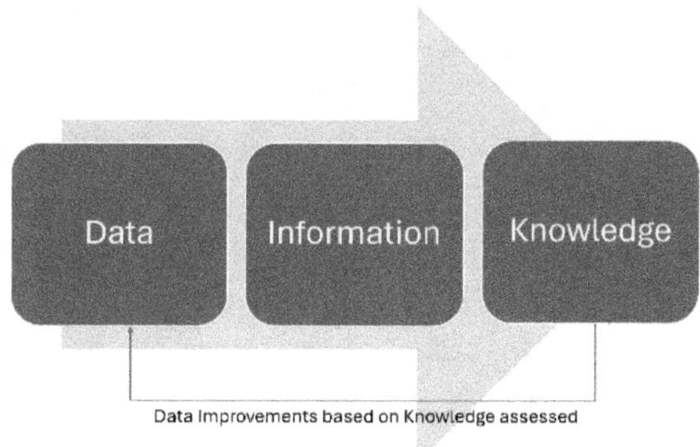

- **Data:** Refers to raw, unprocessed facts or observations. It lacks context and meaning. Examples include temperature readings, stock prices, or a list of names.

- **Information:** Derived from data by organizing, structuring, and interpreting it. It provides context and relevance. Examples include weather forecasts based on temperature data, financial reports, or a summary of customer orders.

- **Knowledge:** A deeper understanding that goes beyond information. It involves insights, patterns, and connections gained through experience, learning, and critical thinking. Examples include understanding the impact of weather patterns on crop yields, recognizing market trends, or solving complex problems.

This progression is iterative, not linear. As we gain more knowledge, we refine our understanding of data and generate better information.

Challenges in Data Processing

Data entering the enterprise is often unstructured, varied, non-standard, and incomplete. Processing this data for decision-making or analysis requires significant manual work, leading to redundant processes, system failures due to data overload, and inconsistent data sources. Enterprises must create complicated business rules and custom processes to transform data into useful information, consuming time and resources and involving multiple staff members, various systems, and even spreadsheets to manage data source handling.

Healthcare Example: Leveraging Fabric

Fabric can bring disparate data streams together into a single database, standardize data elements, convert to the Fast Healthcare Interoperability Resources (FHIR) standard, and then process the data further upstream in the data pipeline.

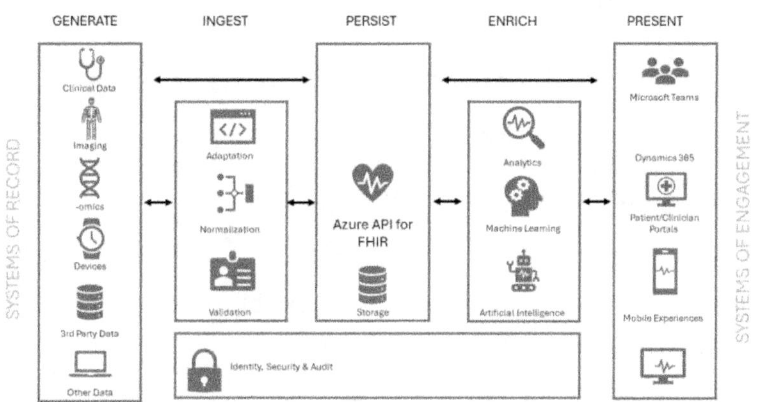

Figure 11 Azure API for FHIR

Leveraging Fabric for the FHIR standard results in:

- A single source of truth.

- Faster processing times.

- Event-triggered processing.

- Scalable infrastructure for growing data needs.

- Insight into performance.

- The ability to deduplicate and link records within and across data types.

This pathway enables more personalized analytics after processes have been completed, leading to more precise reporting and visualizations. By building this data pipeline with separate "Building Blocks" (modular software services that each perform a specific task, which can be combined to create larger data processing and analysis pipelines), we design an architecture that is:

- Easier to understand.

- Flexible and modular.

- Easier to make open-source.

- Consistent in performing similar operations.

Microsoft Fabric is a powerful, centralized analytics platform with the capability to revolutionize how firms leverage their data and AI features. Its architecture comprises three basic layers: the bottom layer is a multi-cloud platform, the middle layer is OneLake data storage, and the top layer is workloads. This innovation provides an inclusive framework that simplifies and rationalizes data administration, enabling organizations to effectively realize the full potential of their data. Its unified approach combines tools in one environment, reducing organizational costs as firms can buy a single pool of computing that drives all Fabric workloads. Therefore, enterprises can leverage their data and AI capabilities while safeguarding customer data security and privacy. Azure OpenAI capabilities are integrated into every Microsoft Fabric layer, and Copilot allows data scientists to use natural language to develop data flows, formulate SQL statements, construct reports, and devise ML models.

Chapter 2:
What is Microsoft Azure

Microsoft Azure is a cloud computing platform that offers a plethora of services without needing your own hardware. With Azure, you can quickly create solutions and accomplish tasks that might be difficult or impossible to handle in your on-premises environment. Azure's comprehensive services for compute, storage, network, and applications enable you to build robust solutions without the burden of managing physical infrastructure.

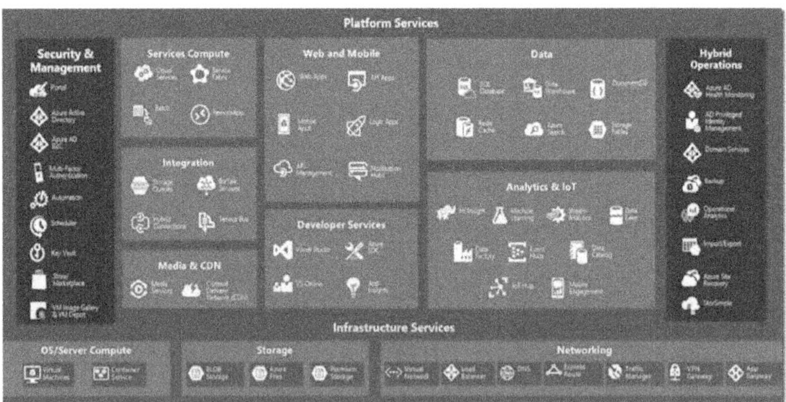

Figure 12 Microsoft Azure Services

Cloud Computing Overview

Operating your own on-site datacenter requires managing every aspect, from purchasing and installing hardware, virtualization, and operating systems, to setting up the network (including wiring), firewalls, and data storage. This process demands significant capital investment for hardware and ongoing operational costs for maintenance. While you have the freedom to choose your hardware and software, you also bear the full cost, regardless of usage levels.

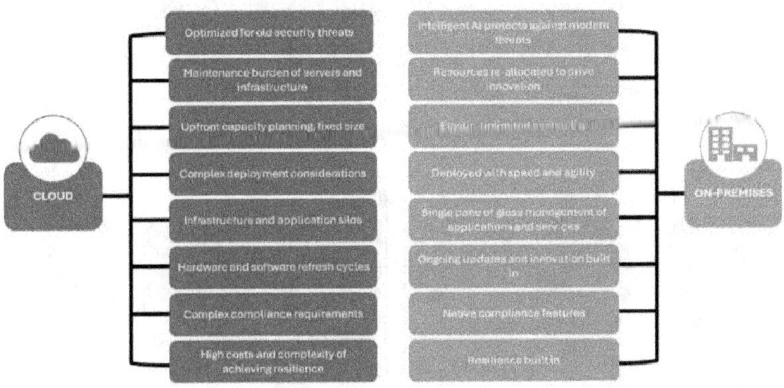

Figure 13 Cloud vs On-prem

In contrast, cloud environments typically offer a web-based portal that allows users to control compute, storage, network, and application resources. For example, in cloud computing, a user can configure a virtual machine (VM) by specifying the size of the compute node (CPU, RAM, and local disks), the operating system, pre-installed software, network configuration, and the node's location. This VM can then be launched and accessed within minutes, a significant improvement over traditional methods that could take weeks just for procurement.

Types of Cloud Environments

Besides public clouds, there are private and hybrid clouds. A private cloud is hosted in your own datacenter and provides self-service access to compute resources within your organization. Although it simulates a public cloud experience, you must purchase and maintain the hardware and software yourself. A hybrid cloud blends public and private clouds, enabling you to host workloads in the most suitable environment. For example, you could run a high-traffic website in the public cloud while connecting it to a secure database in your private cloud.

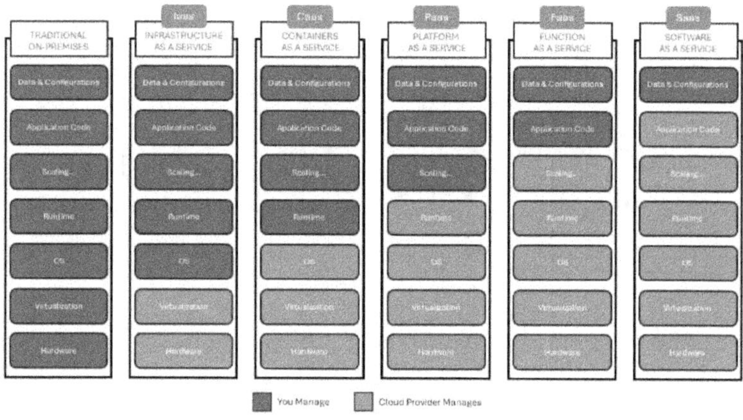

Figure 14 Cloud Computing Models

Microsoft supports various cloud types, including public, private, and hybrid clouds.

Cloud Computing Models

Cloud computing is commonly divided into three categories: Software as a Service (SaaS), Platform as a Service (PaaS), and Infrastructure as a Service (IaaS). These models allow third-party providers to offer services over the cloud, eliminating the need for on-premises data centers. You can access these services on demand via the internet, paying either a subscription fee or on a pay-as-you-go basis.

1. Software as a Service (SaaS) provides software that is hosted and managed by the provider for the customer. In this SaaS model, one application serves all customers and can be scaled for better performance across locations. SaaS products, like Office 365, PowerApps, PowerBI, and more are typically licensed via monthly or annual subscriptions.

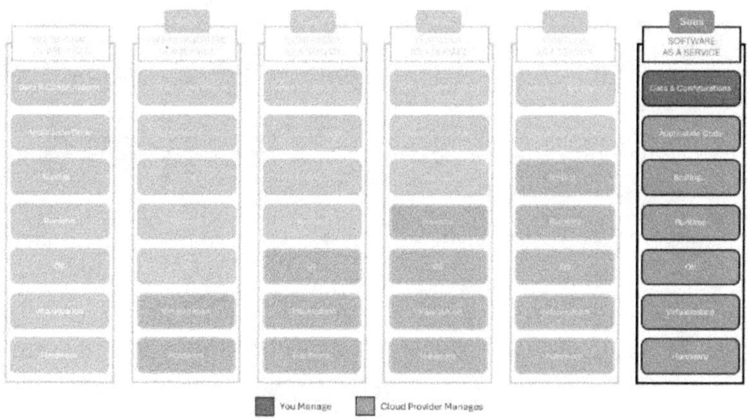

Figure 15 Software as a Service

Subscribers benefit from always having the latest version without the need for manual updates and maintenance. Other examples include Microsoft OneDrive, WordPress, and Dropbox.

2. Platform as a Service (PaaS) allows you to deploy applications to a cloud service vendor's hosting environment. You provide the application, and the PaaS vendor manages its deployment and execution. This frees developers from worrying about the underlying infrastructure.

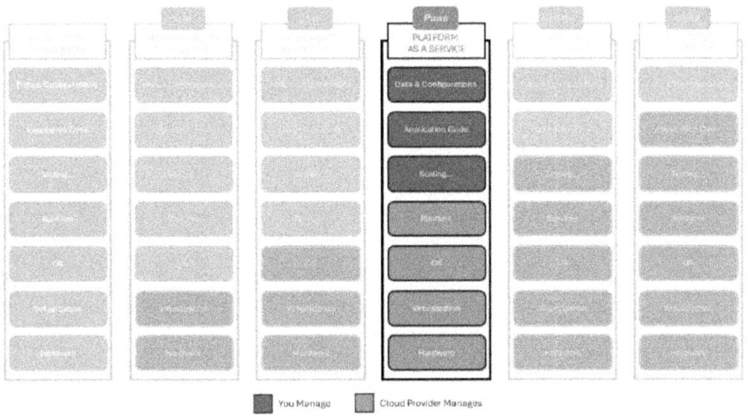

Figure 16 Platform as a Service

Azure offers several PaaS options such as Azure Logic Apps, Azure SQL Database, Azure CosmosDB, and Azure Event Grid where developers can deploy and monitor applications without administering the infrastructure. Scaling and load balancing are handled automatically by Azure, simplifying the process of securing and upgrading these services.

3. Function as a Service (FaaS) offers a platform for creating, executing, and managing application

functions without dealing with the underlying
infrastructure. The cloud service provider
dynamically allocates machine resources, ensuring
efficient operation without the need for manual
intervention.

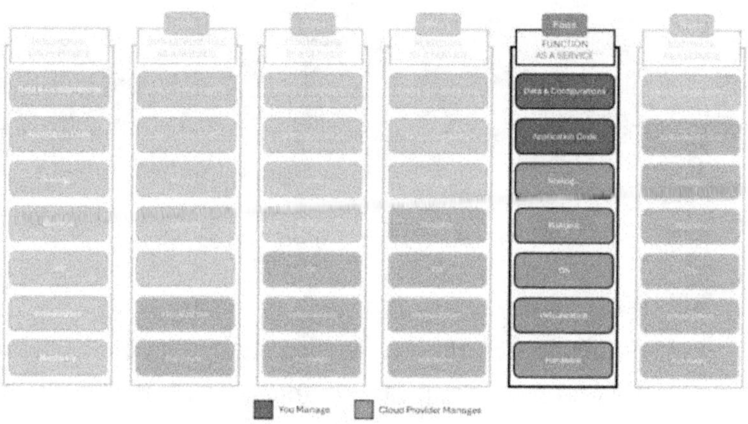

Figure 17 Function as a Service

4. Container as a Service (CaaS) provides the
    necessary hardware and software resources for
    creating and running applications with containers. It
    can be considered a part of or an addition to IaaS,
    focusing on containers instead of VMs. CaaS allows
    developers and IT operations teams to build, launch,
    and manage applications without worrying about
    the infrastructure or platform for running
    containers. The cloud service provider manages the
    environment, while customers remain responsible
    for their code, data, and applications.

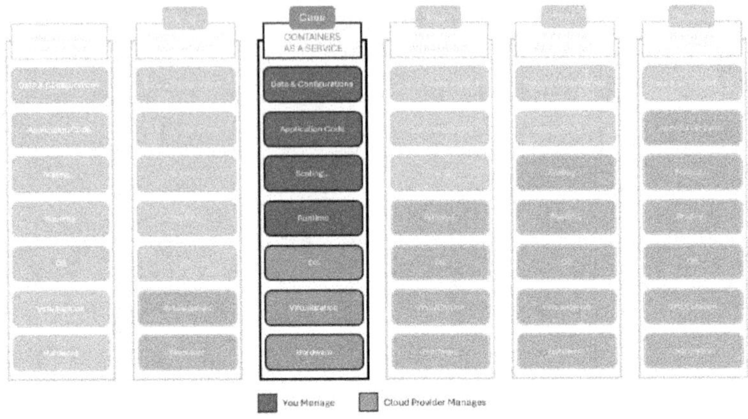

Figure 18 Container as a Service

5. Infrastructure as a Service (IaaS) is where a cloud vendor offering IaaS manages servers with virtualization software, allows you to create virtual machines (VMs) on their infrastructure. You can choose to run either Windows or Linux on your VM and install whatever software you need. Azure provides additional services like virtual networks, load balancers, and storage, all running on its infrastructure. Unlike PaaS, however, you are fully responsible for managing these resources.

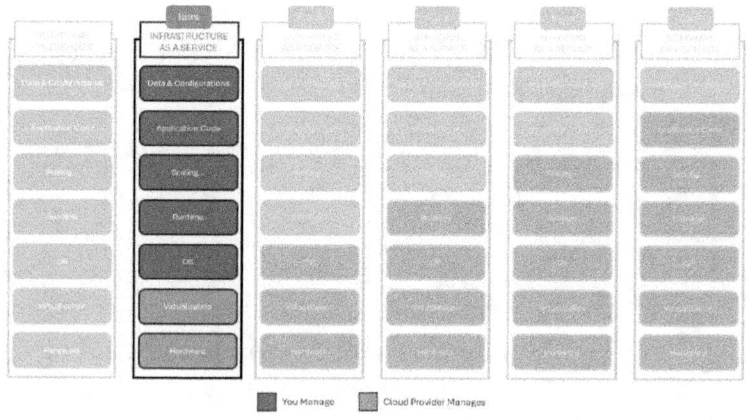

Figure 19 Infrastructure as a Service

Azure Virtual Machines, the IaaS service from Azure, is a popular choice for migrating services to the cloud using the "lift and shift" model. You can set up a VM that mirrors the infrastructure of your current datacenter services and move your software to the new VM. While you may need to make some adjustments, such as updating URLs to other services or storage, many applications can be migrated with minimal changes.

Some key services within Azure's cloud computing platform include:

- **Compute Services:** Azure Cloud Services (web and worker roles), Azure Virtual Machines, Azure Websites, and Azure Mobile Services.

- **Data Services:** Azure Storage (with Blob, Queue, Table, and Azure Files services), Azure SQL Database, and Redis Cache.

- **Application Services:** Azure Active Directory, Service Bus, HDInsight, Azure Scheduler, and Azure Media Services.

- **Network Services:** Azure Virtual Networks, Azure Content Delivery Network, and Azure Traffic Manager.

Microsoft Azure's extensive offerings and flexibility make it a powerful tool for modern cloud computing, enabling businesses to innovate and scale with ease.

Chapter 3: What is a Data Warehouse?

A data warehouse is similar to a database system, but there are some significant differences. Data Warehouses consolidate important data from various databases into a single repository, allowing users to query the warehouse instead of accessing each database individually. This seamless integration hides the complexity of how multiple modules collaborate to process complex queries.

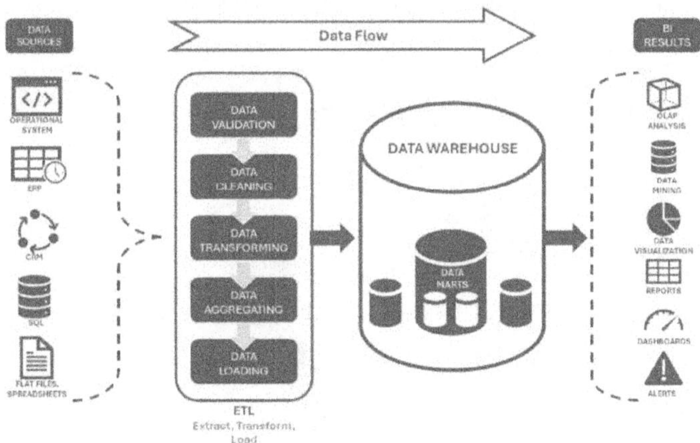

Figure 20 Data Warehouse

A data warehouse is not a specific product but rather an environment designed to gather essential data from diverse databases, often in different formats. This approach addresses the inefficiencies and errors associated with manually collecting and combining data from multiple sources, enabling effective use of complex corporate data, which is a valuable organizational resource.

The primary purpose of a data warehouse is to provide an environment where users can perform queries and analyses without worrying about where the data resides. All queries are sent to the data warehouse as if it's a single database, and the warehouse management system handles query evaluation. By separating analysis and transaction workloads, data warehouses enable organizations to combine and analyze data from different sources, preserving historical records and improving business insights.

A data warehouse environment typically includes:

- An ETL (Extraction, Transformation, and Loading) solution

- Statistical analysis and reporting tools

- Data mining capabilities

- Client analysis tools

- Other applications that manage data collection, transformation, and delivery to business users

Data Warehouse Load Strategies

Loading data into a warehouse is crucial for integrating high-quality, transformed, or processed data into a single repository. Efficient data loading impacts the timeliness and accuracy of analytics, making it essential for real-time decision-making processes. Designing and implementing a robust data loading strategy is vital for the success of a data warehouse project.

In the context of Microsoft Fabric, there are several ways to load data into a warehouse. Proper data loading ensures that high-quality, processed data is integrated into a single repository. Data ingestion involves moving raw data from various sources into a central repository, while data loading focuses on placing transformed data into the final storage destination for analysis and reporting.

OneLake automatically saves all Fabric data items, such as data warehouses and lakehouses, as Delta Parquet files. Staging objects provide temporary storage and transformations, using the same resources as a data warehouse or having their own storage area. This staging area acts as a buffer, minimizing the impact of load operations on the data warehouse's performance, ensuring it remains operational and responsive.

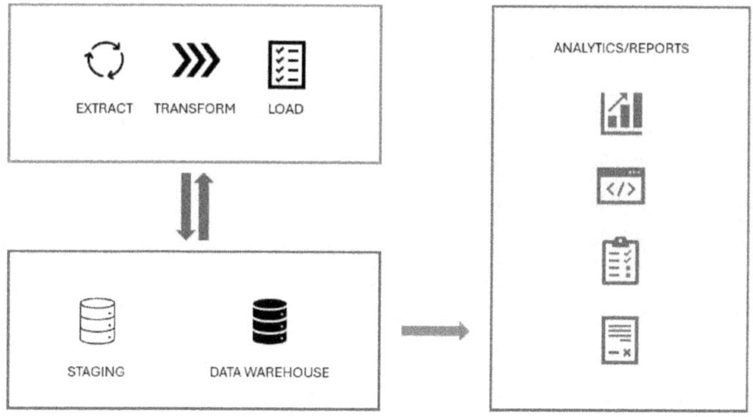

Figure 21 Load Strategies

Data loading can be achieved using multiple strategies, including:

- **T-SQL:** Utilizing the SQL engine for complex queries and data manipulation. Operations involve selecting, ordering, grouping, and merging data from different tables, with various functions and operators for advanced data analysis and transformations.

- **Data Pipelines:** Cloud services for data integration, allowing workflows for data movement and transformation at scale. Complex ETL or ELT processes can be visually created and scheduled, enabling data ingestion and loading from various data stores.

- **Cross Warehouse:** Using the COPY statement to import data into the warehouse. This method offers flexibility in specifying source file formats, designating storage locations for rejected rows, and skipping header rows.

- **Dataflow Gen2:** The next generation of dataflows, providing a complete Power Query experience and simplifying the creation of dataflows, reducing the steps required for data integration.

By implementing these strategies, organizations can ensure efficient, accurate, and timely data loading, enhancing the overall effectiveness of their data warehouse solutions.

Integrating Data into a Data Warehouse

Microsoft Fabric offers a comprehensive platform for data and analytics, featuring enhanced query processing and full transactional T-SQL capabilities for efficient data administration and exploration. Integrating data into a data warehouse involves three main steps: extracting data from various sources, transforming it to meet operational needs, and loading it into the warehouse (ETL). This process ensures that the data is accurate, reliable, and accessible for business intelligence and analytics purposes.

Azure provides several services to facilitate data integration, such as Azure Synapse Analytics, which offers unlimited analytics service with rapid insights, and Azure Data Factory, enabling large-scale hybrid data integration. Recent Azure updates have improved compatibility with Azure Data Lake Store, enhancing real-time analytics on streaming data and providing a robust platform for deploying machine learning models from the cloud to the edge.

Monitoring a Fabric Data Warehouse

A data warehouse is a critical component of a business analytics system. Monitoring it effectively is essential for gaining visibility into its operations. Microsoft Fabric offers several monitoring tools, including:

- **Microsoft Fabric Capacity Metrics App:** Tracks the consumption of capacity units.

- **Dynamic Management Views:** Monitors activities within the data warehouse.

- **Query Insights Views:** Analyzes query patterns.

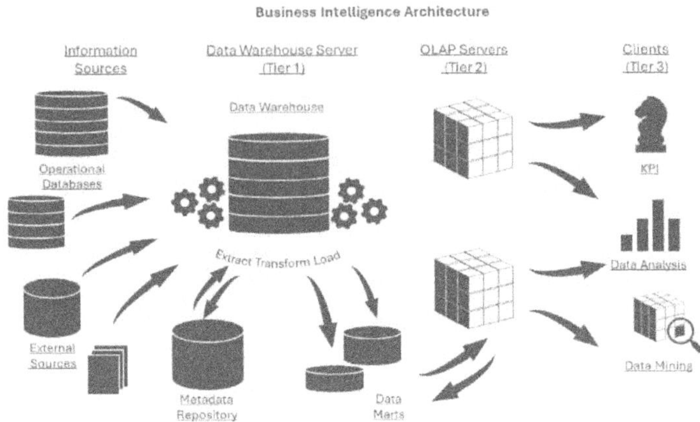

Figure 22 Data Warehouse vs OLTP

Data Warehouse vs. OLTP Systems

A data warehouse is distinct from online transaction processing (OLTP) systems, which handle the daily operations of an organization, such as buying, storing, manufacturing, banking, paying, registering, and accounting. While OLTP systems focus on transaction processing, data warehouses assist users in analyzing data and making decisions. They can organize and present data in various formats to meet different user needs.

Benefits of a Data Warehouse

Data warehousing offers numerous advantages:

- Efficiency: A central location for all essential metrics supports decision-making, reducing the risk of inaccurate reports and enabling data sharing.

Queries and reports enhance management processing efficiency.

- Time Savings: By transforming raw information into valuable analytical tools, data warehouses allow employees to spend more time analyzing information, improving decision-making processes. Traditional OLTP systems can delay information retrieval, slowing down decisions and increasing costs.

- Full Historical Data: Unlike transactional systems that store only current data, data warehouses retain historical data for many years. This enables better business intelligence through time-based analysis, trend analysis, and trend prediction. Historical data also supports data retention requirements for NARA and other governing bodies.

- Improved Data Quality & Consistency: Data from various sources and formats is transformed into a common format, ensuring consistency across departments like sales, marketing, finance, and operations. This builds trust in the organization's data.

Data Warehouses Presentation Layers

- **Information Processing:** Data warehouses enable analysis through queries, statistical methods, crosstabs, tables, charts, and graphs.

- **Analytical Processing:** Tools such as Microsoft SQL Server Analysis Services (SSAS), Oracle Essbase, and IBM Cognos Analytics facilitate multidimensional data analysis. Features like drill-down, roll-up, slicing, dicing, and pivoting help users navigate data hierarchies, supporting strategic decision-making based on data insights.

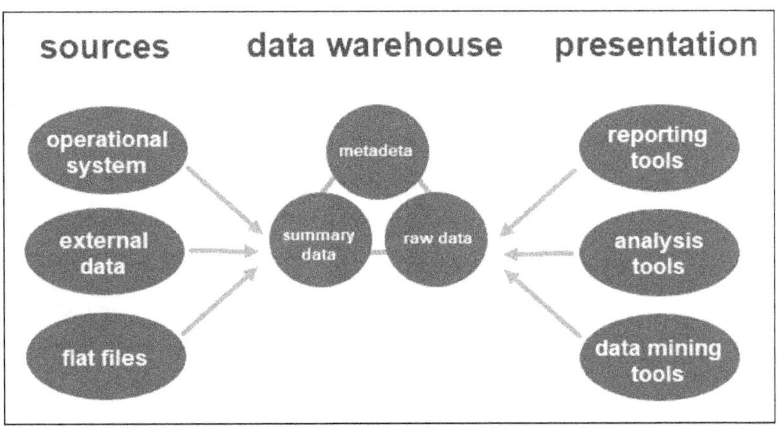

Figure 23 Data Warehouse Layers

Data mining tools analyze and explore data stored in a data warehouse, using sophisticated algorithms to discover patterns and relationships. These tools handle both structured and unstructured data, performing tasks such as clustering, classification, regression, and association. They enable users to predict future trends, identify anomalies, and make data-driven decisions.

Figure 24 Data Mining Steps

Microsoft Fabric supports various data mining tools, including SQL Server Analysis Services (SSAS), which provides an integrated platform for predictive analytics, encompassing data cleansing, preparation, machine learning, and reporting. SSAS includes standard algorithms like clustering, neural networks, and decision trees, and tools for time series forecasting, data mining, and text mining models.

With user-friendly interfaces and visualization capabilities, these tools simplify understanding the results of data mining tasks, playing a crucial role in business intelligence and analytics. They help organizations gain insights from their data and use this information to enhance operations and strategies.

Data Mining Results with Reporting Tools

Data Mining

Data mining helps discover knowledge by uncovering hidden patterns and relationships, building analytical models, and performing classification and prediction. These mining outcomes can be visualized using various tools, making it easier to interpret and act on the findings.

Reporting Tools

A presentation layer data warehouse reporting tool is software that enables users to interact with data in a data warehouse. These tools facilitate the creation, management, and sharing of reports based on the data. They feature easy report design, live data access, and interactive dashboards.

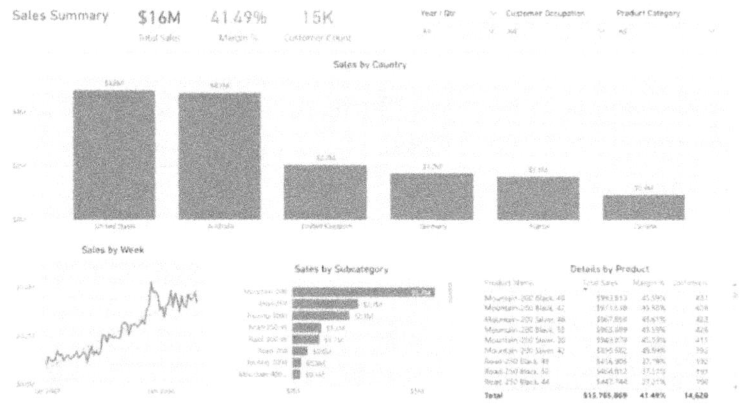

Figure 25 Reporting Example

Microsoft supports the use of multiple reporting tools like Tableau, Power BI, and Looker, which are essential for business intelligence by helping organizations make data-driven decisions. These tools are designed to be user-friendly, allowing even non-technical users to explore complex datasets and gain insights.

Key features of presentation layer data warehouse reporting tools include:

- **Data Visualization:** Present data in various visual formats such as charts, graphs, and tables, making it easier to understand and interpret.

- **Interactive Dashboards:** Provide interactive dashboards that allow users to manipulate data and view it from different perspectives, aiding in trend, pattern, and anomaly identification.

- **Real-Time Access:** Offer real-time access to data, enabling timely decision-making based on the most current information.

- **User-Friendly Interface:** Designed for ease of use, enabling non-technical users to analyze complex datasets and derive insights.

- **Report Generation:** Allow users to create, manage, and distribute reports based on data stored in the warehouse, crucial for sharing insights within the organization.

- **Integration Capabilities:** Integrate with various data sources, allowing for the combination and analysis of data from different systems.

- **Security Features:** Include robust security features to ensure sensitive data is protected and only accessible to authorized users.

Data Lake vs. Data Warehouse

These terms are often confused, but they serve different purposes and are increasingly integrated. Most companies benefit from using both systems. Data warehouses store structured data (organized in rows and columns), while data lakes store all types of data, including structured and unstructured data. This results in a system with higher volume and more complex governance and data management.

DATA WAREHOUSE	Vs.	DATA LAKE
Structured, processed	Data	Structured/semi-structured/unstructured, raw
Schema-on-write	Processing	Schema-on-read
Expensive for large data volumes	Storage	Designed for low-cost storage
Less agile, fixed configuration	Agility	Highly agile, configure and reconfigure as needed
Mature	Security	Maturing
Business professionals	Users	Data scientists et. Al.

Figure 26 Data Warehouse vs Data Lake

Chapter 4: Project Managers' Overview of Fabric

Complete Analytics Platform

Microsoft Fabric's Real-Time Analytics features allow you to ingest, query, and interact with data streams seamlessly. Fabric simplifies scalable analytics, eliminating the need to stitch together services from different providers. It offers a unified, easy-to-learn, configure, build, and manage product. Fabric provides tailored experiences and tools in a unified user interface. Before Fabric (Pre-Fabric) all these capabilities existed in a non-cohesive series of environments.

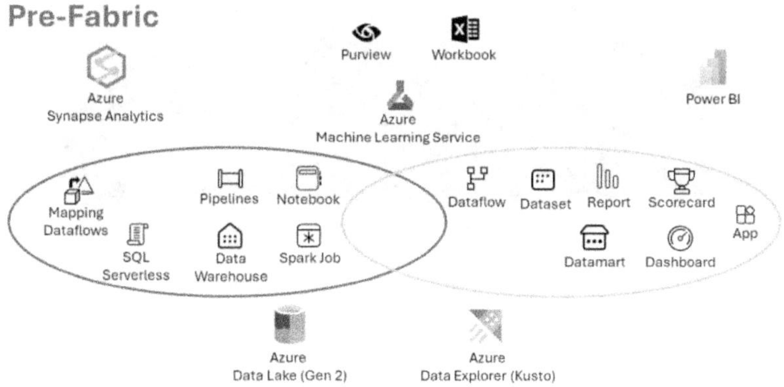

Figure 27 Pre-Fabric

Lake-Centric and Open Architecture

Fabric is a single SaaS platform that enables the use of different analytics engines on data stored in OneLake, an open-format storage solution. It is scalable, affordable, accessible online, and continuously updated and maintained by Microsoft. OneLake merges storage locations from different regions and clouds into one logical lake, similar to OneDrive for data. This architecture avoids the need to transfer and replicate data across systems and teams. OneCopy allows access to data from a single copy without moving or duplicating it. OneLake uses Azure Data Lake Storage (ADLS) as its foundation and supports various data formats such as Delta, Parquet, CSV, and JSON.

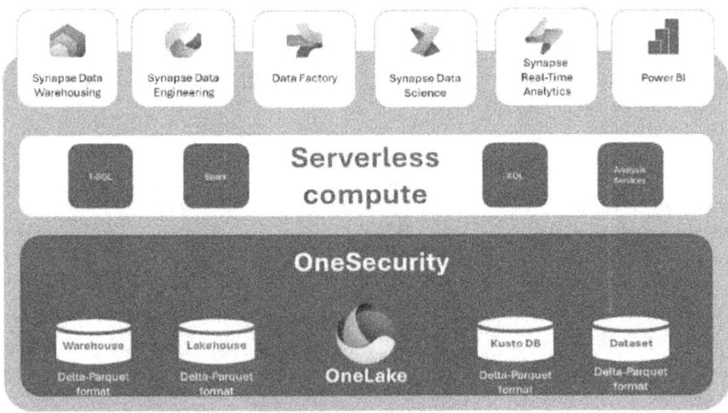

Figure 28 Serverless Compute

Unified Experience for Developers

Fabric helps developers analyze data and present insights to business users through a single interface and architecture, replacing the need for multiple products from different vendors. As a SaaS product, Fabric works efficiently and allows users to start quickly and achieve fast results. It offers different experiences for various roles in the analytics process, including data engineers, data warehouse professionals, data scientists, data analysts, and business users. These experiences include:

- **Power BI:** Connect to data sources, create and explore meaningful visuals, and share them easily. It allows business users to access data quickly and intuitively, improving decision-making.

- **Data Factory:** Provides a modern data integration solution to connect, cleanse, and transform data from various sources using Power Query and over 200 native connectors.

- **Data Activator:** A no-code feature that sets up actions such as email alerts and Power Automate workflows triggered by specific data patterns or conditions.

- **Industry Solutions:** Offers data solutions tailored to different industries, addressing their unique challenges and objectives. Examples include retail, healthcare, and manufacturing solutions that enhance operations and reporting.

- **Synapse Data Engineering:** Provides a Spark platform with excellent authoring capabilities for building, managing, and optimizing data infrastructures.

- **Synapse Data Science:** Enables the creation, execution, and management of machine learning models, integrating with Azure Machine Learning for automatic experiment tracking and model registry.

- **Synapse Data Warehouse:** Delivers industry-leading SQL performance and scale, with independent scaling of storage and compute components, and stores data in the open Delta Lake format.

- **Synapse Real-Time Analytics:** Specializes in streaming and time-series data, offering a cloud-based platform for analyzing and visualizing data as it flows, supporting event-driven data use cases and real-time analytics.

Fabric's suite of tools and unified platform simplifies the analytics process, making it easier for organizations to gain insights and make informed decisions based on their data.

The Power of OneLake

A data lake is foundational for all Fabric workloads, and Microsoft Fabric Lake, or OneLake, is built into the Fabric platform to provide a single location for storing all organizational data. Microsoft OneLake is crucial for IT, offering a platform for data storage and management. This hierarchical data lake is similar to Azure Data Lake Storage (ADLS) Gen2 or the Windows file system.

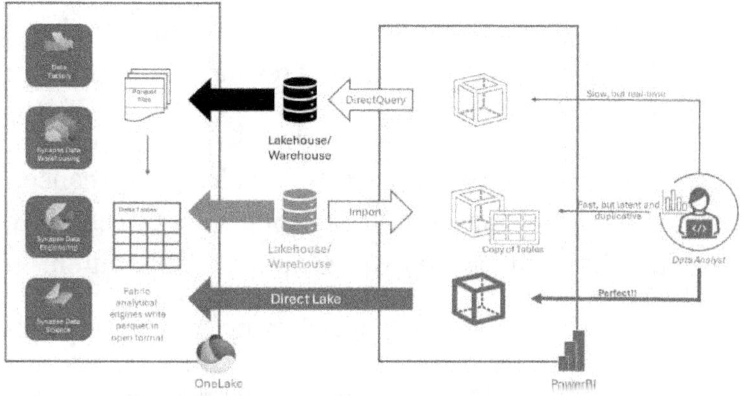

Figure 29 Onelake Overview

OneLake features data access roles, allowing role-based access control (RBAC) for data. You can create security roles that grant read access to specific folders within a Fabric item and assign them to users or groups. These access permissions determine what folders users can view when accessing the lake view of the data through the lakehouse UX, notebooks, or OneLake APIs. This structure enables security settings at different hierarchical levels, effectively managing and securing data in a structured and regulated manner.

One way to envision Microsoft OneLake is as OneDrive for enterprise data as a unified, logical data lake that supports your entire organization. Unlike some other products, you don't need to create separate data lakes for each department; OneLake allows you to store and access all your analytics data in one location.

Benefits of OneLake

1. **Enhanced Collaboration**: OneLake creates a single data lake for the entire organization, reducing the need to manage multiple resources for different business groups.

2. **Governance by Default**: It sets clear governance and compliance boundaries controlled by the tenant admin, with shared ownership for collaboration.

3. **Openness at Every Layer**: Using Azure Data Lake Storage (ADLS) Gen2, OneLake supports any type of file, whether structured or unstructured.

4. **Ease of Use**: OneLake is automatically available with every Microsoft Fabric tenant and requires no additional resources to set up or manage.

OneLake is a single, unified, logical data lake provided by Microsoft Fabric for your entire organization, like OneDrive for enterprise data. Here are the details:

Purpose and Features:

- **Unified Data Lake**: OneLake serves as a central repository for all your analytics data, designed to be the single location for storing and managing data for analytics purposes.

- **Multi-Cloud SaaS Service**: It is the first multi-cloud SaaS data lake for organizations, simplifying data blending, analysis, security, governance, and discovery.

- **Data Duplication Reduction**: By providing a single data lake for the entire organization,

OneLake reduces data duplication and fragmentation.

- **Support for Analytical Engines**: The same data stored in OneLake can be used across multiple analytical engines without needing data movement or duplication.

Key Aspects:

- **Tenant-Based Governance**: Each Fabric tenant has exactly one OneLake, ensuring clear governance boundaries and compliance control.

- **Distributed Ownership**: Workspaces within a tenant allow different parts of the organization to contribute to OneLake without friction.

- **Open and Compatible**: Built on Azure Data Lake Storage (ADLS) Gen2, OneLake supports any type of file (structured or unstructured), storing data in Delta Parquet format.

In conclusion, OneLake provides a unique way to store and manage data, focusing on collaboration, governance, and openness, making it an attractive option compared to other data storage solutions.

Chapter 5:
What is Microsoft Fabric

Microsoft Fabric is a comprehensive data solution platform built on a SaaS cloud approach. It integrates OneLake with various data storage, development, and analytics tools, providing organizations with a unified solution. With the rise of artificial intelligence (AI) and the increasing importance of data science, data has become a critical asset for digital transformation and competitiveness. However, the proliferation of data science tools in the past often led to unnecessary complexities, as experts used a variety of tools to gather, store, analyze, and visualize data. This resulted in disjointed and disorganized DataLakes, complicating their construction, integration, management, and use.

To address these challenges, Microsoft Fabric is positioned as a modern, integrated platform that simplifies the adoption and use of data tools. It offers a comprehensive suite of data services through SaaS, including data transition, processing, analysis, conversion, live event routing, and reporting. Microsoft Fabric centralizes various components, such as Power BI and real-time analytics tools, revolutionizing the data science landscape by incorporating AI.

The Revolutionary Nature of Microsoft Fabric

Microsoft Fabric Architecture

Fabric Architecture (Mohammed, 2023)

Fabric's architecture consists of three primary layers:

1. **Bottom Layer (Multi-Cloud Platform):** This layer supports multiple cloud providers, including Amazon S3, Azure, and Google Cloud, with ongoing integration of additional cloud-based systems. The multi-cloud approach is beneficial for enterprises that rely on more than one cloud solution provider. Users can connect to these cloud solutions, pull data through available "shortcuts," and incorporate it into their workflows.

2. **Middle Layer (OneLake Data Storage Center):** OneLake serves as the centralized data store for Fabric, based on the "data lakehouse" paradigm, which combines the best features of data lakes and data warehouses. It is the main repository for all data within Fabric, transitioning away from relational formats to the 'data lake format.' OneLake is open-source, allowing users to link products that can read from it. One of its key features is the ability to create shortcuts to other data stations, such as AWS S3, eliminating the need for multiple copies of data assets.

3. **Top Layer (Workloads):** This layer includes various workloads such as Azure Synapse Analytics, Power BI, Azure Machine Learning, Data Activator, Synapse Data Warehousing, Synapse Data

Engineering, and Synapse Real-Time Analytics. These workloads provide the tools necessary for advanced data processing, analysis, and reporting.

Key Components of Microsoft Fabric

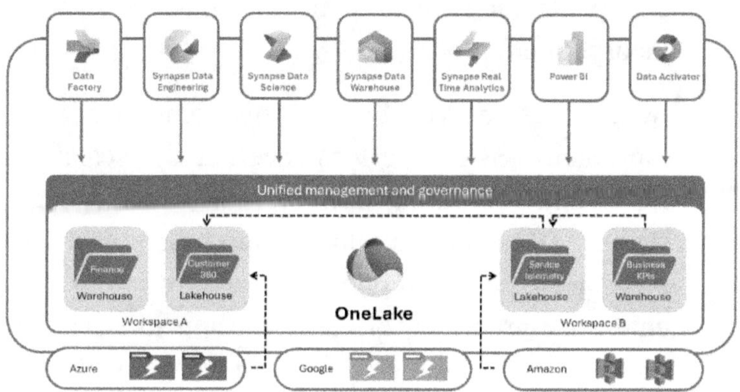

1. **Data Factory:** Data Factory is a modern data integration platform that enables businesses to ingest, prepare, and transform data from a wide range of sources. It simplifies data integration with over 200 native connectors, allowing seamless connectivity with both on-premises and cloud-based data sources.

2. **Synapse Data Engineering:** Synapse Data Engineering provides a Spark service with robust authoring and data transformation capabilities. It enables the creation, organization, and optimization of tools for handling large volumes of data. Its integration with Data Factory allows businesses to schedule and execute notebooks and Spark jobs efficiently.

3. **Synapse Data Science:** Synapse Data Science allows enterprises to build, deploy, and manage machine learning (ML) models directly from Fabric. It integrates with Azure ML for experiment tracking and model registry, enabling data scientists to enrich enterprise data with predictive insights, which can be incorporated into BI reports.

4. **Synapse Data Warehouse:** This component offers top-notch SQL performance and scalability, separating computing from storage for independent scaling. It introduces a lake-centric data warehouse built on a distributed processing platform, providing enterprise-grade performance without the need for complex configuration and management. The warehouse seamlessly integrates with Power BI for easy analysis and reporting, merging the worlds of data lakes and warehouses.

5. **Synapse Real-Time Analytics (SRTA):** SRTA is a big data analytics platform optimized for streaming and time-series data. It reduces data complexity and simplifies integration, allowing businesses to gain real-time insights and visualize their data in motion, all while maintaining powerful analytical capabilities.

6. **Power BI:** Power BI is a versatile business analytics tool within Fabric that allows businesses to connect to their data sources, visualize, and share insights. It supports interactive dashboards and includes many inbuilt data connectors for a seamless experience.

Power BI consists of a desktop application, a SaaS service, and mobile apps for iOS and Android.

7. **Data Activator:** Data Activator is a zero-code feature in Fabric that allows users to detect specific patterns in their continuously evolving data and trigger actions, such as email alerts. It evaluates data in Power BI reports and event streams, taking recommended actions when certain thresholds or patterns are met.

In summary, Microsoft Fabric is a transformative platform that unifies various data tools and services, making it easier for organizations to manage, analyze, and derive insights from their data. Its architecture and integrated components are designed to streamline data processes, enhance scalability, and support advanced analytics, ultimately driving better business decisions and outcomes.

Microsoft Fabric Implications

Unification

Microsoft Fabric is transforming the data science landscape by bringing together previously disjointed data components and services into a unified platform. Built on a SaaS foundation, it seamlessly integrates both new and existing tools such as Data Factory and Data Activator into a single environment. Each tool is tailored for specific user roles, such as data engineers and warehousing professionals, creating a cohesive experience. By leveraging AI, Fabric accelerates data movement, enabling users to access a wide array of deeply connected analytics tools. This unified approach also offers the advantage of quick access to and easy reutilization of all assets, with centralized data lake storage preserving data in its original form while allowing the use of the customer's chosen analytics tools. Microsoft Fabric thus ensures seamless organization, governance, administration, and discovery of data and services, while also providing robust security features for assets, data, and access.

Cost Reduction

The integration of processes within Microsoft Fabric leads to significant cost reductions. Traditional analytics systems often require products from multiple vendors for a single project, resulting in the use of computing capacity across different systems. When one system becomes idle, its unused capacity cannot be leveraged by others, leading to inefficiencies. Microsoft Fabric addresses this issue by allowing businesses to purchase and manage a single pool of computing resources that power all Fabric workloads. This comprehensive approach enables users to design solutions that utilize all workloads without any difficulty, reducing costs as unused capacity in one workload can be repurposed for others.

Empowering Business Users

Microsoft Fabric empowers business users by facilitating data-driven decision-making. By integrating Fabric with Microsoft 365 apps commonly used in businesses, Microsoft fosters a data culture within organizations. Workloads such as Power BI have already been embedded into Microsoft 365, allowing users to seamlessly discover data from OneLake and extract more value from it. For example, employees can use Excel to access and analyze data in OneLake and then generate Power BI reports, turning Microsoft 365 apps into powerful hubs for uncovering and applying insights.

Acceleration of AI Adoption and Deployment

Microsoft Fabric integrates Azure OpenAI capabilities into every layer, enabling data professionals to accomplish more with less effort. The Copilot feature allows data scientists to use natural language to create data flows, generate SQL statements, build reports, and develop machine learning models. It also allows users to create custom conversational experiences by integrating Azure OpenAI workflows with their data, which can then be published as plug-ins. This AI-driven approach empowers businesses to harness the power of AI, even if they lack direct expertise in AI or data science.

The impact of Microsoft Fabric on Generative AI (GenAI) and AI, in general, is profound. It aids in data standardization and facilitates data access, making it easier for businesses to use AI without the need to move or copy data. Microsoft Fabric's universal system allows its data lake to incorporate shortcuts that point to other storage locations, enabling GenAI to access data more easily. Even if the stored data is in different formats, Fabric can standardize, troubleshoot, and organize it to make it compatible with GenAI applications. Additionally, as various GenAI apps extract insights from Fabric and introduce new data into the system, Fabric ensures synchronization.

Microsoft Fabric is a groundbreaking, centralized analytics platform that has the potential to revolutionize how organizations harness the power of their data and AI capabilities. Its architecture, comprising a multi-cloud platform, OneLake data storage center, and various workloads, provides a comprehensive framework that simplifies and streamlines data management. By unifying tools in a single environment, Microsoft Fabric reduces organizational costs and enhances the ability of enterprises to leverage their data and AI while ensuring the security and privacy of customer data.

Chapter 6: Core Workloads in Microsoft Fabric

In the chapters above we have had an overview of the seven main workloads that make up Microsoft Fabric: Data Factory, Synapse Data Engineering, Synapse Data Science, Synapse Data Warehouse, Synapse Real Time Analytics, Power BI, and Data Activator. We shall explore more deeply into each of these core workloads in this chapter.

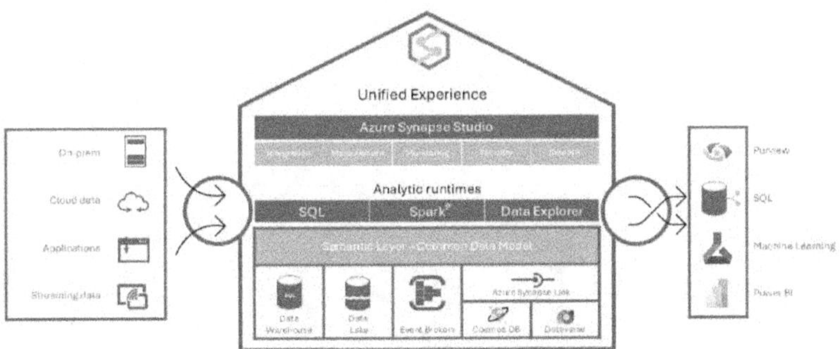

Figure 30 Unified Fabric Experience

Data Factory

Data Factory provides modern data integration capabilities to gather, organize, and transform data from various sources, including real-time data, Lakehouse, data warehouses, and databases. This service allows both individual users and developers to manipulate data with intelligent transformations and use a rich set of activities (Kromer et al., 2023). Data Factory also offers fast copy features to data pipelines and dataflows in Microsoft Fabric (Kromer et al., 2023). This rapid data movement enables easy transfer between preferred sources and destinations. For example, users can swiftly move their data to a data warehouse and organize it in Microsoft Fabric for relevant analytics. The primary functions of Data Factory include dataflows and pipelines, which support over 300 transformations in the dataflow designer, enhancing flexibility and transformation. Conversely, data pipelines facilitate rich data orchestration to suit flexible data workflows.

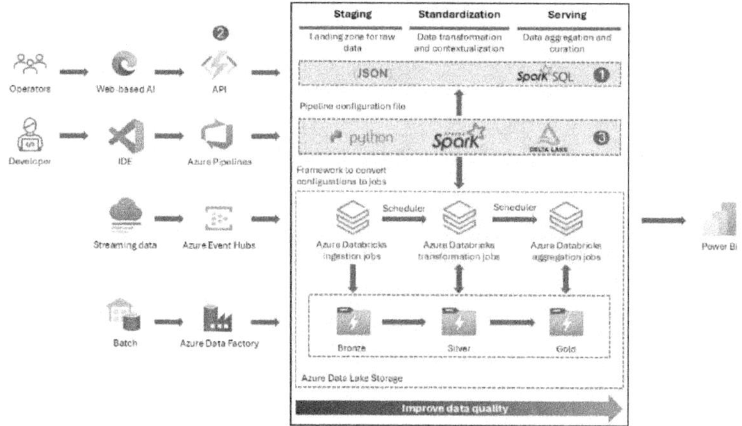

Figure 31 **Azure Flow**

Pipelines

Dataflows Gen2 provide an excellent option for data transformations in Microsoft Fabric. When additional operations are required on transformed data, dataflows and pipelines can be used together.

Data pipelines are easily created in the Data Factory and data engineering workloads. Pipelines are a common concept in data engineering and offer a wide variety of activities to orchestrate. Some common activities include:

- Visualizing activities in a specific order.

- Using a dataflow for data ingestion and transformation, and landing into a Lakehouse using dataflows.

- Incorporating the dataflow into a pipeline to orchestrate additional activities, such as executing scripts or stored procedures after the dataflow is complete.

Dataflows

The process of collecting data is essential for analytics. With Microsoft Fabric's Data Factory, Dataflows enable multi-step data collection and transformation visually using Power Query Online.

Dataflows are a type of cloud-based ETL (Extract, Transform, Load) tool for building and executing scalable data transformation processes.

Dataflows Gen2 allow the extraction of data from various sources, transformation using a wide range of operations, and loading into a destination. Power Query Online offers a visual interface for these tasks.

Fundamentally, a dataflow includes all transformations to reduce data prep time, and can be loaded into a new table, included in a Data Pipeline, or used as a data source by data analysts.

Traditionally, data engineers spend significant time extracting, transforming, and loading data into a consumable format for downstream analytics. The goal of Dataflows Gen2 is to provide an easy, reusable way to perform ETL tasks using Power Query Online.

If only using a Data Pipeline, data is copied, and preferred coding languages are used to extract, transform, and load it. Alternatively, a Dataflow Gen2 can be created first to extract and transform data. Data can also be loaded into a Lakehouse and other destinations, making it easy for the business to consume the curated semantic model.

Adding a data destination to your dataflow is optional, preserving all transformation steps. To perform other tasks or load data to a different destination after transformation, create a Data Pipeline and add the Dataflow Gen2 activity to your orchestration.

Another option might be to use a Data Pipeline and Dataflow Gen2 for an ELT (Extract, Load, Transform) process. For this order, a Pipeline extracts and loads data into a preferred destination, such as the Lakehouse. A Dataflow Gen2 then connects to the Lakehouse data to cleanse and transform it. In this case, the Dataflow is offered as a curated semantic model for data analysts to develop reports.

Dataflows can be horizontally partitioned. Once a global dataflow is created, data analysts can use dataflows to create specialized semantic models for specific needs.

Dataflows promote reusable ETL logic, reducing the need to create multiple connections to data sources. They offer a wide variety of transformations and can be run manually, on a refresh schedule, or as part of a Data Pipeline orchestration.

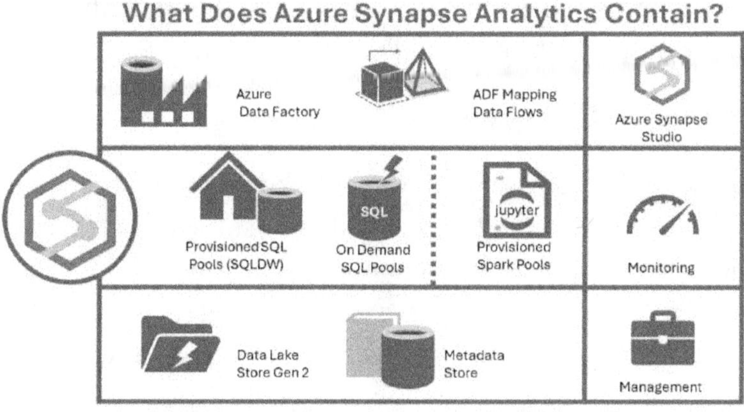

Figure 32 Synapse Data Engineering

Synapse Data Engineering

Synapse Data Engineering is designed to work efficiently with large amounts of data. For example, users can create a Lakehouse for all their organizational data (Lucznik, 2023). In these scenarios, the data engineering service effectively combines the data lake and warehouse to reduce the complexity of ingesting, transforming, and sharing data. The Lakehouse is also designed as a primary workspace item, enhancing availability. Moreover, Synapse Data Engineering provides optimal runtime with strong admin controls and high default performance (Lucznik, 2023). The service's public version includes Runtime with Python, Delta, and Spark. The Spark Runtime also connects to the entire Microsoft Fabric workspace, aiding developers in using the service. Additionally, Synapse Data Engineering's main authoring canvas is Microsoft's Notebook, offering native Lakehouse integration (Lucznik, 2023). Data engineers can also use R and Python libraries or Data Wrangler for low-code scenarios. These features enable easy Spark authoring, collaboration capabilities, and quick start options.

Lakehouse Overview

Microsoft Fabric is based on a Lakehouse architecture, which leverages OneLake's scalable storage layer and Apache Spark and SQL computation engines for big data processing. A Lakehouse is a consolidated platform that integrates the adaptable and scalable storage of a data lake with the ability to query and analyze data typical of a data warehouse.

Common Use Cases

Common scenarios involve a company that has been relying on a data warehouse to store structured data from operational systems, such as order history, inventory levels, and customer information. Additionally, they've gathered unstructured data from social media, website logs, and external sources that are difficult to handle and analyze using the current data warehouse infrastructure. To enhance decision-making abilities by analyzing data in different formats across multiple sources, the company has selected Microsoft Fabric.

Lakehouse Structure

A Lakehouse is a database-like structure that uses Delta format tables on a data lake foundation. Lakehouses merge the relational data warehouse's ability to analyze data with SQL and the data lake's adaptability and scalability. They can handle any data format and work with various analytics tools and programming languages. As cloud-based solutions, lakehouses can dynamically adjust their size and offer high reliability and backup.

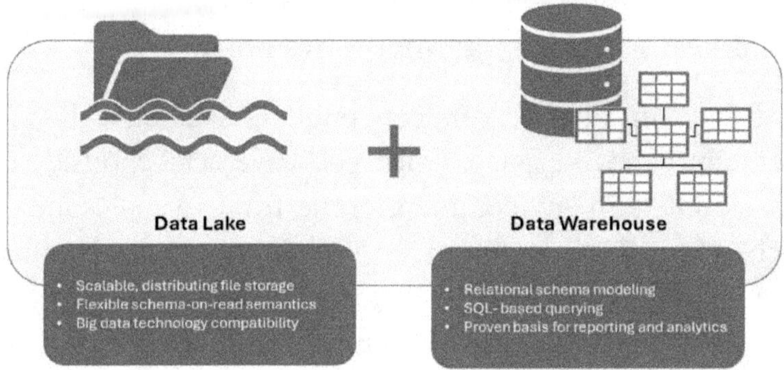

Figure 33 Datalake vs Data Warehouse

Benefits of a Lakehouse

- **Integration of Spark and SQL Engines:** Lakehouses combine Spark and SQL engines to handle large data volumes and perform machine learning or predictive analytics.

- **Schema-on-Read Format:** Data is stored in a schema-on-read format, meaning the schema is defined when needed instead of being predefined.

- **ACID Transactions:** Lakehouses ensure ACID (Atomicity, Consistency, Isolation, Durability) transactions with Delta Lake tables for data consistency and integrity.

- **Collaborative Workspace:** Lakehouses provide a common workspace for data engineers, data scientists, and data analysts to collaborate on data.

A Lakehouse is an ideal choice if you need a scalable analytics solution that maintains data consistency. It's essential to assess your specific needs to decide which solution is most appropriate.

Building a Lakehouse in Microsoft Fabric

In Microsoft Fabric, you can build a lakehouse in any premium-tier workspace. Once you have a lakehouse, you can load data in any common format from various sources, such as local files, databases, or APIs. You can also automate data ingestion using Data Factory Pipelines or Dataflows (Gen2) in Microsoft Fabric. Furthermore, you can create Fabric shortcuts to data in external sources like Azure Data Lake Store Gen2 or a Microsoft OneLake location outside the lakehouse's own storage. The Lakehouse Explorer allows you to navigate files, folders, shortcuts, and tables, and view their contents within the Fabric platform. After loading data into the Lakehouse, you can use Notebooks or Dataflows (Gen2) to examine and transform it. Data Factory Pipelines can orchestrate Spark, Dataflow, and other activities, enabling you to implement complex data transformation processes.

Data Management and Analysis

After transforming data in the Lakehouse, you can query, train, analyze, or report on it. You can also manage data policies, such as data classification and access control.

There are several methods to load data into a Fabric:

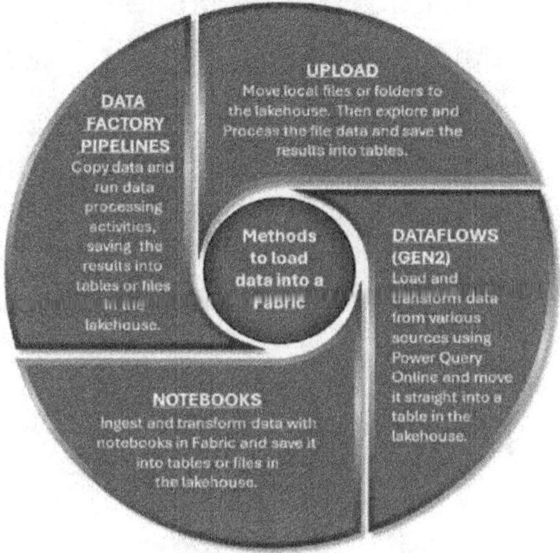

Figure 34 Loading Data into Fabric

Fabric Shortcuts

Shortcuts allow you to incorporate data into your lakehouse without moving it from external storage. They are helpful when sourcing data from different storage accounts or even different cloud providers. Within your Lakehouse, you can set up shortcuts that refer to different storage accounts and other Fabric items such as data warehouses, KQL databases, and other Lakehouses. OneLake handles source data permissions and credentials. To access data with a shortcut to another OneLake location, the user's identity is used to check their access to the data in the shortcut's target path. The user needs read permissions in the target location. Shortcuts can be created in Lakehouses and KQL databases and appear like a folder in the lake, allowing Spark, SQL,

Real-Time Analytics, and Analysis Services to query the data.

Lakehouse Medallion Architecture

Today's business world relies on comprehending and using vast and diverse data sources. The Fabric Lakehouse, which merges data lakes and data warehouses, provides an excellent platform to process and explore this data. The medallion architecture is a common industry standard for analytics based on the Lakehouse. It helps manage and enhance your Lakehouse environment. The medallion architecture consists of three levels: bronze, silver, and gold, which help you organize, refine, and verify your data.

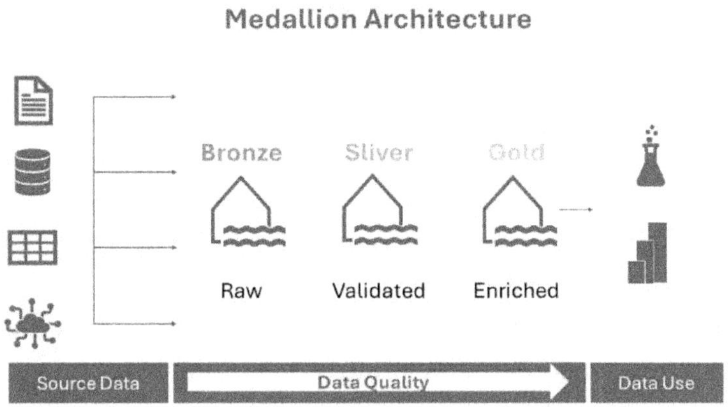

Figure 35 Medallion Architecture

Bronze Layer

The first layer of the Lakehouse is the bronze or raw layer. It's where all data is ingested, regardless of whether it is structured, semi-structured, or unstructured. The data retains its original format and remains unprocessed.

Silver Layer

The second layer is the silver or validated layer, where data is checked and improved. Common tasks in this layer include joining and merging data and applying data validation rules such as removing nulls and duplicates. The silver layer can be seen as a central store for an organization or team, where data is stored uniformly and can be used by multiple teams. Here, data is cleaned and prepared for further refinement and modeling in the gold layer.

Gold Layer

The third layer is the gold or enriched layer. In this layer, data is further refined to meet specific business and analytics needs. This might include aggregating data to a particular level, such as daily or hourly, or adding more information from external sources. When the data reaches the gold layer, downstream teams, like analytics, data science, or MLOps, can use it.

Customizable Medallion Architecture

Medallion Architecture offers flexibility to adapt to your organization's specific needs. It allows for customization of layers based on your requirements. For example, you may need an additional "raw" layer to handle data in a specific format before processing it into the bronze layer. Conversely, a "platinum" layer might be necessary for advanced data processing and enhancement for specialized purposes. The architecture's flexibility means that both the names and number of layers can be adjusted as needed.

Data Transformation vs. Data Orchestration

Data Transformation involves modifying data formats or content to meet specific needs. Fabric provides several tools for data transformation:

- **Dataflows (Gen2)**: Suitable for smaller semantic models and straightforward transformations. They allow for basic data formatting and transformations.

- **Notebooks**: Best for complex transformations and larger semantic models. Notebooks support detailed data manipulations and can store transformed data as managed Delta tables in the lakehouse, prepared for reporting.

Data Orchestration entails managing and coordinating data-related processes to achieve specific outcomes. Fabric's primary tool for data orchestration is:

- **Pipelines**: A sequence of steps that transfers data between different layers of the medallion architecture. Pipelines can be configured to run automatically based on schedules or triggered by specific events, ensuring smooth data flow and processing.

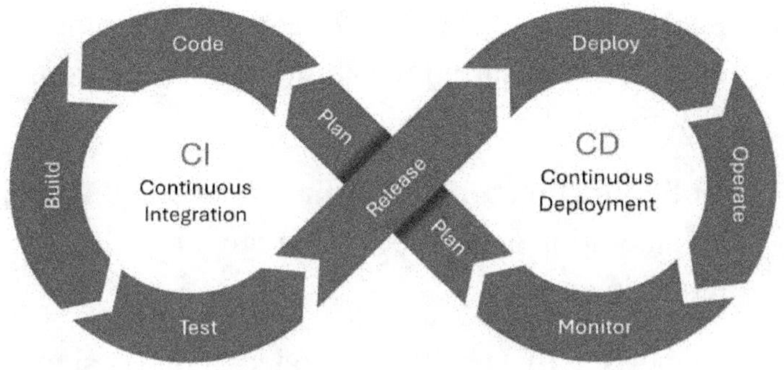

Figure 36 CI/CD Flow

Leveraging CI/CD Processes

Implementing Continuous Integration/Continuous Deployment (CI/CD) in a lakehouse architecture involves several critical factors to ensure effective and smooth deployment:

- **Data Quality Checks**: Ensuring the accuracy and consistency of data throughout the pipeline.

- **Version Control**: Managing changes to data and code to track modifications and revert if necessary.

- **Automated Deployments**: Streamlining the deployment process to minimize manual intervention and reduce errors.

- **Monitoring and Security**: Continuously overseeing the pipeline's performance and securing data against unauthorized access.

- **Scalability and Disaster Recovery**: Planning for growth and ensuring the ability to recover from failures.

- **Collaboration and Compliance**: Facilitating team collaboration and adhering to regulatory requirements.

Fabric supports CI/CD with integrated tools designed to streamline these processes. Key components include:

- **Azure Repos and Azure Pipelines**: Developers push code changes to Azure Repos branches. This action triggers Azure Pipelines to compile, test, and build artifacts, which are then deployed into production environments.

- **GitHub Actions**: Allows the creation of custom CI/CD workflows, providing additional flexibility and automation for various development tasks.

CI/CD pipelines play a crucial role, especially at the gold layer of a lakehouse, where they ensure that high-quality, validated, and reliable data is available for use. Automated CI/CD processes facilitate the continuous integration of new data and transformations, minimizing manual errors and providing consistent, up-to-date insights. This enhances data accuracy, accelerates decision-making, and supports data-driven initiatives effectively.

By leveraging Fabric's CI/CD capabilities, organizations can improve scalability, security, and overall quality, ensuring that their data pipelines remain efficient and dependable.

Securing the Lakehouse

You can protect your lakehouse data from unauthorized access using Fabric. Fabric allows you to manage permissions at both the workspace and item levels.

Workspace Permissions: Control access to all items within a workspace. This is useful for managing access among team members in the same workspace.

Item-Level Permissions: Manage access to specific items within a workspace. This is ideal for cases where users need access to only one particular item or are outside the primary workspace.

By organizing different layers of your lakehouse into separate workspaces, you can enhance security and manage capacity more efficiently. This approach improves both security and cost-effectiveness.

Security and Access Requirements:

- **Bronze Layer**: Restrict access to read-only to minimize exposure.

- **Silver Layer**: Decide if users should have permissions to create new items, balancing flexibility and security.

- **Gold Layer**: Limit access to read-only to ensure data integrity and confidentiality.

Apache Spark in Microsoft Fabric

Apache Spark is an open-source framework designed for large-scale data processing and analytics. It supports parallel data processing across multiple nodes, making it ideal for big data scenarios.

Microsoft Fabric enables you to run Spark clusters, facilitating efficient data processing within a lakehouse environment. Spark operates by distributing data processing tasks across a cluster of computers, using a driver program and SparkContext for task management.

Spark supports multiple programming languages including Java, Scala, R, Spark SQL, and PySpark (Python for Spark). PySpark and Spark SQL are commonly used for data engineering and analytics tasks.

Spark clusters in Microsoft Fabric come with popular libraries pre-installed. To add more libraries or customize your environment, you need workspace admin permissions. You can also save library specifications by creating and setting up environments as defaults for your workspace.

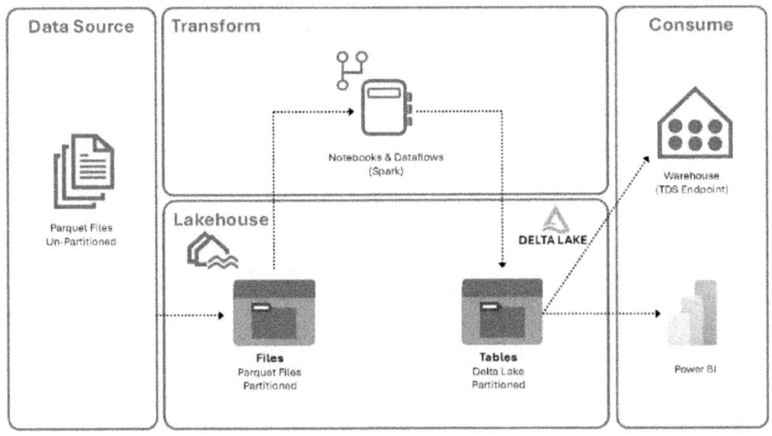

Figure 37 Parquet Files

For data storage, the Parquet format is recommended due to its efficiency and compatibility with large-scale data analytics systems. Converting data from formats like CSV to Parquet often simplifies data transformation needs.

Notebooks in Microsoft Fabric support basic charting features for visualizing data queries. For more advanced visualizations, you can use Python graphics libraries integrated within the notebooks.

Creating Data Visualizations and Leveraging Machine Learning

Data Visualizations with Python Graphics Packages

Visualizing data through code can be highly effective, and Python offers a wealth of graphics packages for this purpose. Most of these libraries build on the powerful Matplotlib framework. Using these graphics libraries, you can embed visualizations directly within notebooks, seamlessly integrating code for data processing with inline visualizations and explanatory markdown cells. This combination enhances both the analysis and presentation of your data.

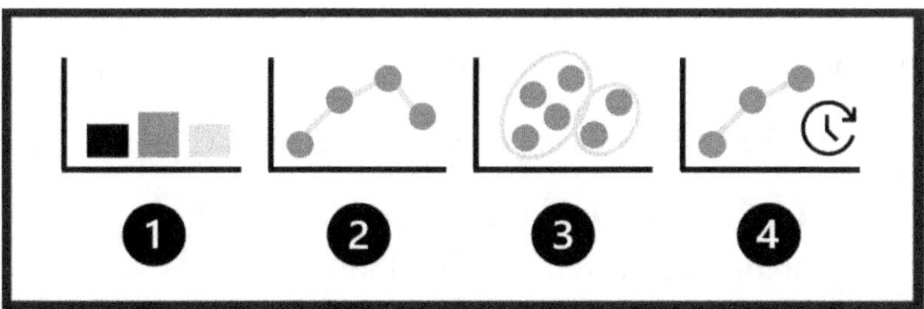

Figure 38 Regression Models

Synapse Data Science provides a comprehensive, end-to-end workflow for developing, collaborating on, and managing complex AI models. It integrates data science with business intelligence and analytics within Microsoft Fabric (Gustafsson, 2023). Key features include:

- **Data Wrangler**: Simplifies data preparation and Python code generation.

- **Synapse ML Library**: Facilitates scalable and distributed machine learning with user-friendly APIs.

- **PREDICT Function**: Enables scalable predictions without moving data.

- **R Language Support**: Enhances versatility with universal usability.

Unlocking Insights with Machine Learning

Machine learning empowers you to uncover patterns within your data and make informed predictions. For example, you can forecast next week's product sales based on historical trends. Machine learning models analyze data to reveal insights that drive strategic decisions.

Here are the four primary types of machine learning models:

- **1. Classification**: Categorizes data, such as predicting customer churn likelihood.

- **2. Regression**: Estimates continuous values, like predicting product costs.

- **3. Clustering**: Groups similar data points, revealing natural data segments.

- **4. Forecasting**: Projects future values from time-series data, such as estimating next month's sales.

Before selecting a model, it's crucial to understand the business problem and the available data.

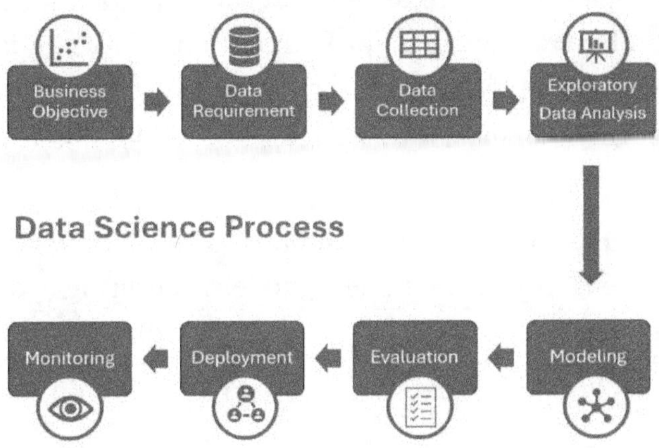

Figure 39 Data Science Process

The Data Science Process

1. **Establish the Goal**: Collaborate with business users and analysts to define the model's objectives and success metrics.

2. **Acquire the Data**: Identify and access data sources, storing them in a Lakehouse.

3. **Process the Data**: Import data from the Lakehouse into a notebook, clean, and transform it as needed.

4. **Train the Model**: Choose an appropriate algorithm and hyperparameters, experimenting and tracking results with MLflow.

5. **Produce Insights**: Use batch scoring to generate predictions and insights.

As a data scientist, the majority of your efforts will focus on data processing and model training. The quality of your data handling and algorithm choice significantly impacts the success of your model.

Tools and Libraries

Utilize open-source libraries to prepare and train models in your preferred language. For Python, consider using Pandas and Numpy for data processing, and libraries like Scikit-Learn, PyTorch, or SynapseML for model building.

Track and manage your experiments with MLflow in Microsoft Fabric. This tool simplifies the monitoring and deployment of your trained models, ensuring you can manage performance and make data-driven decisions effectively.

Elevating Data Science with Microsoft Fabric

Optimizing Data for Machine Learning

Data science thrives on high-quality, extensive datasets, especially when training machine learning models for artificial intelligence. The performance of models typically improves with larger datasets, but data quality remains equally critical.

Microsoft Fabric's robust data ingestion and processing capabilities ensure both the quality and quantity of your data. You can leverage either a low-code or code-first approach to set up data ingestion, exploration, and transformation pipelines. Once your data is ingested, explored, and preprocessed, you can train your model-an iterative process that requires diligent monitoring.

Tracking and Logging with MLflow

Microsoft Fabric integrates with MLflow to track and log your machine learning experiments. This allows you to review and optimize your training processes effectively. Each experiment in MLflow can consist of multiple runs, each representing a task such as model training. Artifacts, including model metadata, are saved within these runs, making your results reproducible.

You can save these artifacts as registered models in Microsoft Fabric, with each new model version tracked under the same name. This streamlined approach simplifies model management and version control.

Leveraging Microsoft Fabric Notebooks

Microsoft Fabric notebooks are a powerful tool for exploring data and discovering patterns. They facilitate easy integration of your findings into a data science workflow and enable you to generate reports using tools like Power BI. Notebooks combine code, text, and visualizations in a single interactive document, enhancing your ability to prototype and share insights efficiently.

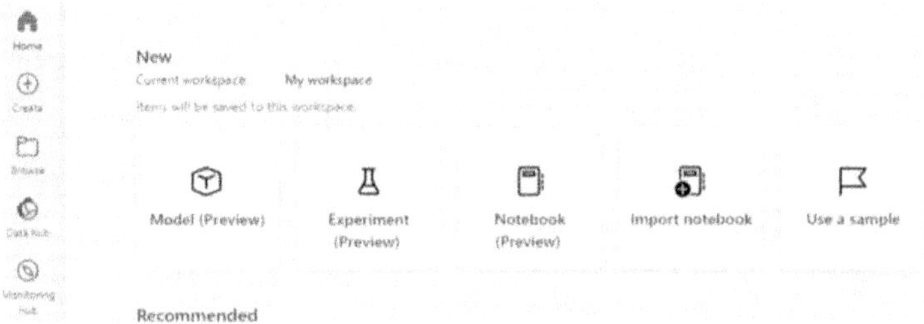

Figure 40 Fabric Notebooks

Data scientists benefit from the interactive nature of notebooks, which support:

- **PySpark (Python)**

PySpark is the **Python API for Apache Spark**. It allows real-time, large-scale data processing in a distributed environment using Python. It also includes a PySpark shell for interactive data analysis.

- **Spark (Scala)**

Apache **Spark** is an analytics engine designed for large-scale data processing. It offers high-level APIs in Java, **Scala**, Python, and R, along with an optimized engine that supports general execution graphs.

- **Spark SQL**

Spark SQL is a **module for structured data processing** that introduces the concept of DataFrames, providing a programming abstraction and functioning as a distributed SQL query engine.

- **SparkR**

SparkR is an **R package** offering a lightweight interface to use Apache Spark from R. It also supports distributed machine learning via MLlib.

Collaborative Notebook Features

Microsoft Fabric notebooks offer real-time collaboration, enhancing efficiency in data science projects. Multiple users can simultaneously edit the same document, which is invaluable for pair programming, remote troubleshooting, and knowledge sharing. For example, if you encounter a coding issue, you can invite a colleague to review and assist within the same notebook. Real-time updates, including cursor movements and code modifications, provide instant feedback and accelerate problem-solving.

Microsoft Fabric notebooks empower data scientists to work together seamlessly, solving problems and learning collectively in real time.

Streamlining Data Management with Microsoft Fabric

Effortless Data Handling with Notebooks

Microsoft Fabric notebooks transform data management by providing a seamless interface for interacting with your data assets. Once your data is stored in the lakehouse, you can easily create code within notebooks to access and utilize these assets, simplifying the process of data consumption and manipulation.

Harnessing Data Wrangler's Power

Data Wrangler in Microsoft Fabric revolutionizes data preparation with its intuitive visual interface. It allows you to explore and preprocess data efficiently, ensuring it's clean and ready for machine learning models. Key features include:

- **Dynamic Data Display**: Instantly updates with each operation, showing summary statistics and visualizations.

- **Code Generation**: Automatically produces reusable code for data cleaning and preprocessing tasks.

- **Preprocessing Tools**: Offers a suite of functions for data cleaning, feature engineering, and exploration.

Data Wrangler enhances data quality by providing a graphical interface for examining and transforming data.

This results in more accurate models and better predictions by addressing issues such as missing values, outliers, and incorrect data types.

Elevating Data Quality for Machine Learning

Effective data preprocessing is critical for building precise machine learning models. Data Wrangler's grid-style data display and dynamic summary statistics enable thorough exploration and feature engineering. By resolving data quality issues and preparing data comprehensively, you set the stage for more reliable model training and insightful predictions.

Introducing Synapse Data Warehousing

Synapse Data Warehousing integrates lakehouse and data warehouse functionalities, supporting open data formats natively (Sathy, 2023). This innovation allows IT teams to collaborate effectively while maintaining security and governance. It leverages SQL for multi-table ACID transactions, utilizing the SQL Server Query Optimizer and Distributed Query Processing engine. Key benefits include:

- **Full Integration and Optimization**: Offers self-optimization, auto-scaling, and cross-querying capabilities.

- **Simplified Management**: New warehouses can be created with ease, including naming and sensitivity labeling.

- **Efficient Data Loading**: Data can be loaded using T-SQL queries with the COPY command.

Revolutionizing Data Warehousing

Data warehouses, which support SQL queries through a relational schema, are simplified in Microsoft Fabric. Tables in a Fabric lakehouse use the Delta Lake format, an open-source storage layer that enhances Spark's capabilities with relational database features. Delta Lake integrates seamlessly with Apache Spark, combining the benefits of a relational database with the flexibility of a data lake. This architecture enables robust SQL-based data operations with transactions and schema validation, creating a powerful analytical data store.

Maximizing Data Insights with Semantic Models and Delta Lake

Unlocking Data Relationships with Semantic Models

A semantic model visually maps how tables interrelate, groups, and aggregates data, and utilizes calculations to reveal insights. In Power BI, these relationships and metrics form the foundation for comprehensive reporting. Fabric automatically generates a semantic model whenever a data warehouse is built, making it readily available for analysts and business users to craft detailed reports.

Seamless Integration with Power BI

With Fabric, the semantic model is seamlessly integrated into the reporting process. If you're familiar with Power BI, working with these models will be straightforward. The data warehouse updates the semantic models automatically, reducing maintenance efforts. However, you can also create custom models tailored to your specific needs.

Automatic and Custom Semantic Models

Fabric provides a default semantic model that aligns with the business logic of your lakehouse or warehouse. This model is continuously updated and synchronized, simplifying your analytics workflow. It includes new tables from the lakehouse by default but also allows you to manually add tables or views for enhanced flexibility.

Advanced Analytics with Delta Lake

Delta Lake enhances data processing by combining the robustness of relational databases with the scalability of data lakes. Here are key benefits of Delta tables:

- **Dynamic Data Management**: Delta tables support CRUD operations, allowing you to query, add, modify, and delete data similarly to relational databases.

- **ACID Transactions**: Delta Lake ensures reliable data modifications with atomicity, consistency, isolation, and durability, thanks to its transaction log and serializable isolation.

- **Versioning and Time Travel**: The transaction log tracks all changes, enabling data versioning and time travel to access historical data.

- **Batch and Streaming Data**: Delta Lake supports both batch and streaming data, making it versatile for various data processing needs with Spark Structured Streaming.

- **Compatibility and Formats**: Data in Delta tables is stored in Parquet format, a standard for data lakes, and can be queried using SQL through Microsoft Fabric's analytics endpoint.

By leveraging semantic models and Delta Lake's capabilities, you can transform raw data into actionable insights, streamline your analytics processes, and drive more informed decision-making.

Microsoft OneLake: Centralizing Your Data Ecosystem

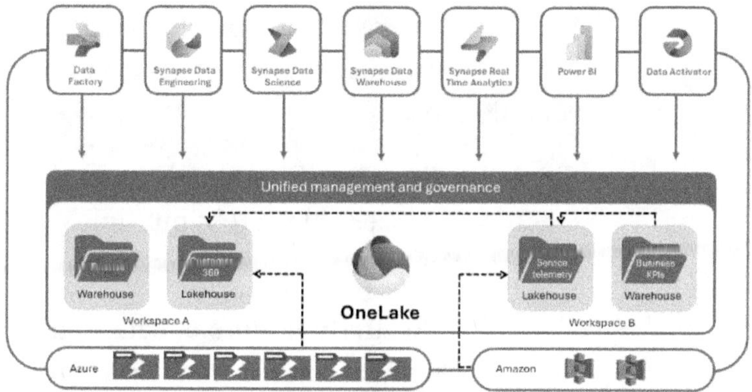

Figure 41 OneLake

OneLake: The Unified Data Lake

Microsoft OneLake is the cornerstone of Microsoft Fabric, designed around the concept of a unified data lake. Here's what this entails:

- **Integrated Data Lake**: OneLake offers a single, cohesive data lake for the entire organization, similar to how OneDrive centralizes document storage. It's automatically available to every Microsoft Fabric tenant with no additional setup or management required.

- **Streamlined Data Management**: By consolidating data into one storage location, OneLake reduces data fragmentation, minimizes duplication, and simplifies data movement. This eliminates the need for multiple isolated data lakes, thereby enhancing efficiency and reducing complexity.

Core Features:

- o **Unified Storage**: OneLake integrates all data from various projects and users into one central repository.

- o **Cross-Engine Data Sharing**: Data in OneLake is accessible across multiple analytical engines.

- o **Enhanced Security**: Future updates will include integrated security models to protect data within OneLake.

Merging Lakehouse and Data Warehouse Capabilities

Seamless Integration with SQL Analytics Endpoint

In Microsoft Fabric, the Lakehouse automatically creates a Warehouse, and the SQL analytics endpoint enables efficient querying of data:

- **Automatic SQL Endpoint**: Each Lakehouse comes with its own SQL analytics endpoint, allowing the use of T-SQL and the TDS protocol to query data. This endpoint presents Delta tables as SQL tables, ensuring compatibility with SQL queries.

- **Automatic Metadata Discovery**: The SQL analytics endpoint automatically updates SQL metadata, including statistics, by reading Delta logs. This ensures up-to-date information without manual data import or copying. The endpoint leverages the same engine as the Warehouse for rapid and low-latency SQL queries.

- **Versatile Analytics Scenarios**: The integration of Lakehouse and Data Warehouse enables flexible analytics, supporting batch, streaming, and lambda architecture patterns. For example, you can analyze data from the gold layer of a medallion architecture using the SQL analytics endpoint, even if the data is stored outside of Microsoft Fabric's OneLake.

Streamlined Data Pipelines with Microsoft Fabric

Efficient Data Coordination

Microsoft Fabric simplifies data ingestion and transformation through Data Factory pipelines:

- **Pipeline Functionality**: Data pipelines orchestrate a series of actions to move and transform data from multiple sources to target destinations. They are crucial for automating extract, transform, and load (ETL) processes, moving data from operational stores to analytical data stores like Lakehouse or Data Warehouse.

- **Familiar Structure**: If you're experienced with Azure Data Factory, you'll find Microsoft Fabric's pipelines familiar. They follow the same structure of linked activities to set up complex data processing tasks and control flow logic.

- **On-Demand and Automated Execution**: Pipelines in Microsoft Fabric can be run manually via the user interface or set to execute automatically, offering flexibility in managing data workflows.

- **Activity Sequencing**: Pipelines consist of a series of activities that define data transfer and transformation tasks. Activities can be linked in a specific order, allowing for complex processing logic such as branching and looping. The outcome of each activity - whether success, failure, or completion - determines the flow to subsequent activities.

- **Custom and Predefined Pipelines**: You have the flexibility to create custom pipelines tailored to your data processing needs. Microsoft Fabric also offers predefined pipeline templates for common scenarios, which can be customized to fit specific requirements.

- **Validation and Execution**: Once a pipeline is designed, you can use the Validate option to ensure its configuration is correct. After validation, pipelines can be executed on demand or scheduled for automated runs.

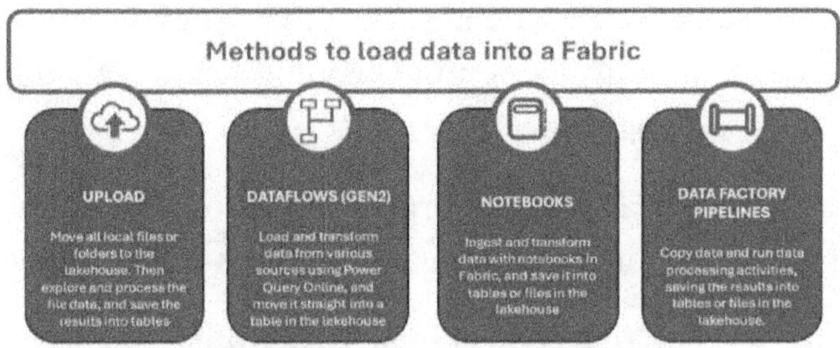

Figure 42 Loading Data into Fabric

Microsoft Fabric simplifies data ingestion with several visual-based tools and methods:

- **Dataflows**: Dataflows allow for a code-free approach to data preparation, cleaning, and transformation. Using Power Query Online, you can graphically design multi-step data ingestion processes, modifying data types, adding or removing columns, and creating calculated columns.

- **Cross-Warehouse Ingestion**: This feature enables the creation of new tables or the merging of tables from different sources within the same workspace. Utilize Transact-SQL features such as INSERT...SELECT, SELECT INTO, or CREATE TABLE AS SELECT (CTAS) for effective cross-warehouse data operations.

- **Choosing the Right Tool**:

    - **COPY (Transact-SQL)**: Ideal for high-throughput operations or integrating data within Transact-SQL logic.

o **Data Pipelines**: Best for workflows requiring minimal coding for reliable data ingestion.

o **Dataflows**: Perfect for custom transformations with a code-free approach.

o **Cross-Warehouse Ingestion**: Useful for creating new tables or merging data within the same workspace.

Synapse Real-Time Analytics

Harnessing Real-Time Data Insights

Microsoft Fabric excels in real-time analytics with Synapse Real-Time Analytics, leveraging KQL Database and Kusto Query Language (KQL) for powerful data analysis:

- **KQL Database**: Facilitates table storage and analysis using KQL, ideal for uncovering insights and patterns from both structured and text data, including time-series data from logs or streams.

- **Scalability and Accessibility**: Synapse Real-Time Analytics allows for scalable analytics solutions, ensuring data accessibility across your organization. It requires only a source and a Kusto database to enable data streaming and querying with KQL.

- **KQL Querysets**: Microsoft Fabric supports the distribution of KQL Querysets within the tenant and to Power BI through the quick create feature, enhancing data accessibility and sharing.

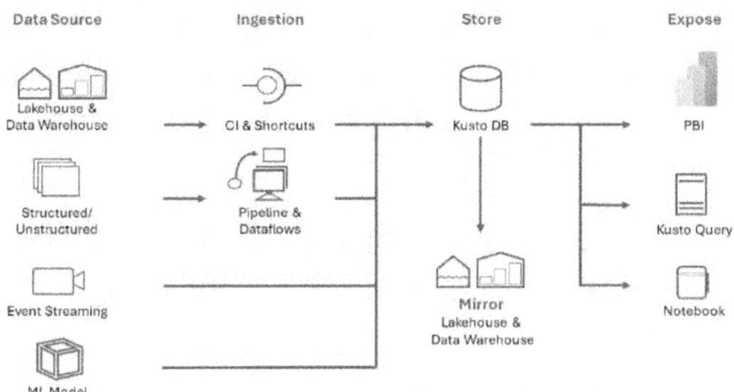

Figure 43 Flow into Kusto DB

Synapse Real-Time Analytics within Microsoft Fabric revolutionizes how IT teams manage and analyze massive volumes of data from logs, telemetry, IoT devices, and more. Leveraging a powerful query language and engine, it seamlessly handles unstructured, semi-structured, and structured data (Schuster et al., 2023). Fully integrated with all Fabric products, it simplifies data loading, transformation, and visualization, while supporting unlimited data and concurrent users with built-in auto-scaling that dynamically adjusts resources based on workload factors such as ingestion rates, CPU usage, and memory demands.

Unlike traditional systems, Synapse Real-Time Analytics empowers data engineers to execute analytical queries directly on raw data without the need for scripting or complex data models (Schuster et al., 2023). This streamlines analytical performance and reduces overall costs, making it an ideal choice for high-speed, large-scale data analysis.

Key Benefits of Synapse Real-Time Analytics:

- **Flexible Data Handling:** Seamlessly ingest and process data from any source and in any format, including structured, unstructured, and semi-structured data.

- **Direct Query Execution:** Perform rapid queries on raw data without complex transformations or models.

- **Optimized Streaming:** Automatically stream and index data for near-instantaneous analysis, with default partitioning by time and hash for efficiency.

- **Scalability:** Efficiently manage data ranging from gigabytes to petabytes, with no limits on concurrent queries or users.

- **Integration:** Work effortlessly with all other Microsoft Fabric features, from data loading to advanced visualization.

Synapse Real-Time Analytics excels in handling time-series data and is particularly effective for scenarios involving IoT and log analytics across diverse industries, including manufacturing, healthcare, retail, and agriculture. It offers a managed service experience with high efficiency, rapid response times, and broad operator support, ensuring that you can derive actionable insights from your data quickly and reliably.

Kusto Query Language (KQL)

Kusto Query Language (KQL) is a powerful, declarative query language designed for rapid and efficient search of large-scale log data. Ideal for cloud-based data analytics, KQL excels in uncovering insights, detecting anomalies, identifying patterns, and creating statistical models. It's employed in Azure Monitor Logs, Azure Application Insights, Azure Resource Explorer, and more, enabling deep and comprehensive data exploration.

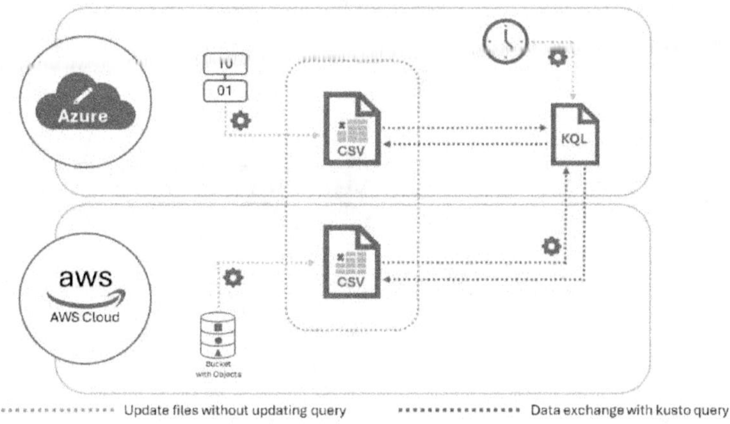

Figure 44 KQL looking at Multi-Cloud log files

Core Components of Real-Time Analytics:

- **KQL Database:** A specialized database in Kusto that includes tables, stored functions, materialized views, shortcuts, and datastreams for robust data management.

- **KQL Queryset:** Allows you to run, view, manipulate, and share query results. Save and export queries, leveraging KQL's capabilities and T-SQL compatibility for diverse querying needs.

- **Eventstream Feature:** Streams data from various sources like Event Hubs and custom apps, delivering it to destinations such as Lakehouse or KQL Database for real-time data handling.

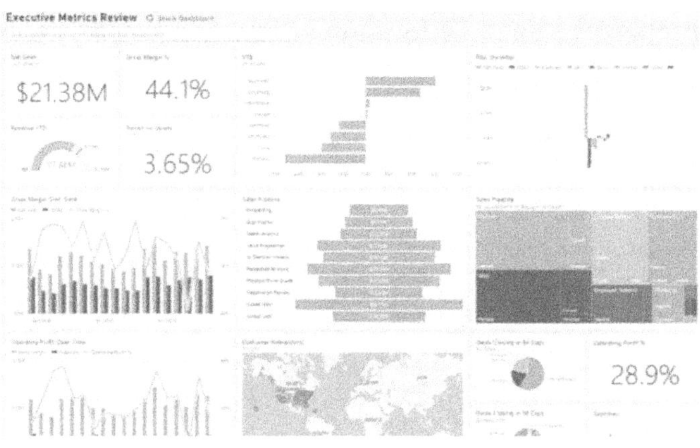

Figure 45 Power BI Dashboard

Power BI within Microsoft Fabric transforms how users interact with data from diverse sources, including on-premise warehouses, cloud-based repositories, and Excel sheets. Key features include:

- **Copilot Integration:** Introduces next-generation AI to Power BI, enhancing data analysis with large language models. Copilot simplifies tasks like DAX calculations and report personalization, delivering insights rapidly.

- **Unified Data Integration:** Power BI's Direct Lake and OneLake modes prevent vendor lock-in and minimize data duplication, ensuring seamless data management across the platform.

- **Advanced Collaboration:** Features top-tier Git integration for linking with Azure DevOps repositories. Track changes, restore versions, and manage team updates efficiently.

Reports **in Power BI:** A Power BI report aggregates visuals from a semantic model to illustrate data relationships and insights. Reports can range from single visuals to multi-page presentations. Users interact with these visuals to explore and analyze data based on their roles and permissions.

Reports vs. Dashboards: While both dashboards and reports feature visualizations, they serve different purposes. Dashboards offer a high-level overview, while reports provide detailed, interactive data exploration. Understanding their distinctions helps in effectively leveraging their unique strengths.

Capability	Dashboards	Reports
Pages	One page	One or more pages
Data sources	One or more reports or semantic model per dashboard	A single semantic model per report
Filtering	No, you can't filter or slice	Yes, there are many ways to filter, highlight, and slice
Can see underlying semantic model tables and fields	No. Can export data but cant see the semantic model tables and fields in the dashboard itself	Yes. Can see semantic model tables and fields and values that have permissions to see
Customization	No	Yes, depending on your permissions you can cross-filter, change visual type, apply design features, add bookmarks and comments, generate QR codes, analyze in Excel, and much more
Pinning	Can pin existing visuals (tiles) only from current dashboard to your other dashboards	Can pin visuals (as tiles) and entire report pages to any of your dashboards

Figure 46 Dashboards vs Reports

Datasets in Power BI

A **Power BI dataset** is the foundation for reporting and visualization within the Power BI environment. It represents a collection of data either imported into or connected from various sources, such as databases, Excel files, or APIs. Datasets enable you to consolidate data from multiple sources into a single location, facilitating comprehensive analysis and discovery.

Dataset Modes:

- **External Datasets:** Import data from external sources into Power BI to create visuals and reports. This mode allows you to analyze static data by integrating it into the Power BI environment.

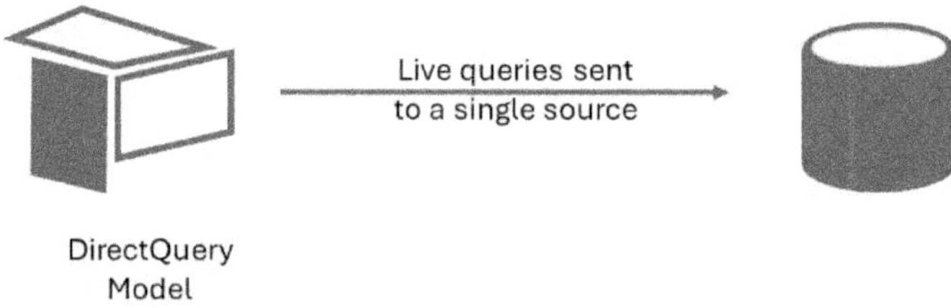

Figure 47 Direct Query Model

- **DirectQuery Datasets:** Link directly to a data source without importing the data. Power BI queries the data in real-time from the source system, ensuring that visualizations reflect the most current information.

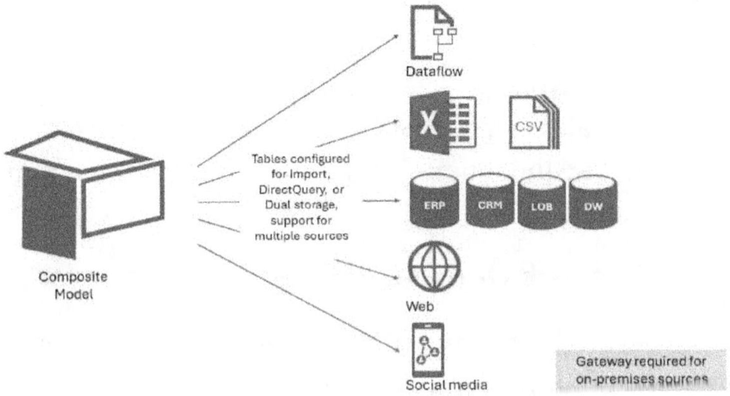

Figure 48 Composite Model

- **Datasets:** Combine and blend data from both imported datasets and DirectQuery connections. Create relationships between tables to enrich your analysis.

- **Dataflows:** Modify, clean, and format data before incorporating it into a dataset. Dataflows streamline the ETL (Extract, Transform, Load) process, preparing data for analysis.

- **Dataset-Workspace Link:** Connect multiple workspaces to a single dataset, fostering collaboration and data sharing across teams.

Data Model and Connections: Design and enhance your data model within a dataset by setting up tables, columns, relationships, calculated columns, measures, and hierarchies.

Direct Lake Mode is a semantic model feature in Power BI that handles massive data volumes with exceptional efficiency. It reads parquet-formatted files directly from a data lake, bypassing the need for a lakehouse or warehouse endpoint, and avoiding data importation into Power BI. This mode facilitates rapid data access and analysis while mitigating the limitations of traditional data import and query methods.

Comparison with Other Modes:

- **DirectQuery Mode:** Retrieves data directly from the original source, reflecting changes immediately but potentially experiencing slower performance due to data retrieval time.

- **Import Mode:** Improves performance by preloading data for DAX and MDX reports, but requires periodic refreshes to reflect source changes and involves data duplication.

Direct Lake Mode Advantages:

- **Real-Time Updates:** Access data without the need for import, reflecting changes as they occur.

- **Enhanced Performance:** Combines the efficiency of import mode with the real-time data retrieval of DirectQuery.

- **Large Data Handling:** Ideal for analyzing extensive and frequently updated models.

Security Features: Direct Lake mode ensures robust data security through row-level and object-level access controls, maintaining the integrity and confidentiality of data.

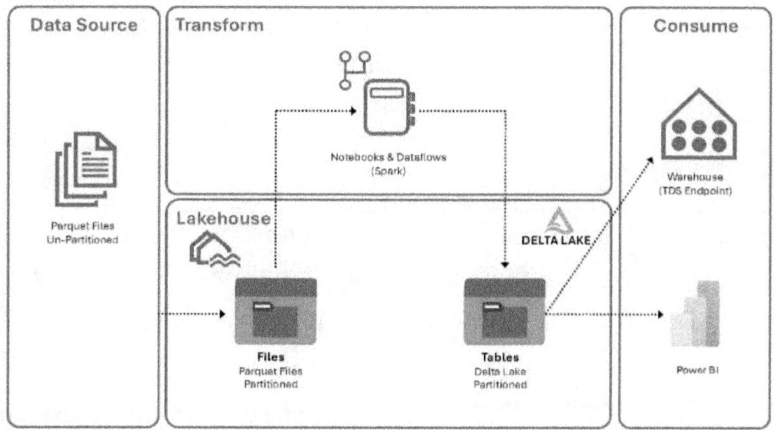

Figure 49 Lakehouse and Direct Lake

Lakehouse

To leverage Direct Lake, you must set up a lakehouse (or warehouse) equipped with at least one Delta table within a supported Microsoft Fabric workspace. The lakehouse is essential as it stores your parquet-formatted files in OneLake and provides access to the Web modeling feature for building Direct Lake models.

Setting Up a Lakehouse: For detailed instructions on setting up a lakehouse, creating a Delta table, and building a basic model, refer to the guide on creating a lakehouse for Direct Lake.

SQL Endpoint

Upon creating a lakehouse, a SQL endpoint is automatically established for executing SQL queries and generating a default model for reports. This endpoint is crucial when switching a Direct Lake model to DirectQuery mode, especially if the data source requires advanced features like security or custom views that Direct Lake cannot interpret. Additionally, Direct Lake uses the SQL endpoint to manage schema and security information.

Data Warehouse

Alternatively, you can use a data warehouse to manage tables through SQL commands or data pipelines. Setting up a separate data warehouse follows a process similar to that of a lakehouse.

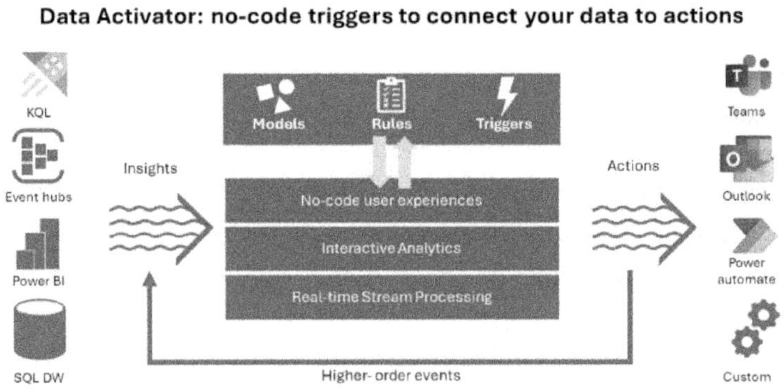

Figure 50 Data Activator

Data Activator

Microsoft Fabric's Data Activator revolutionizes how you manage your data by automating actions without coding. This service continuously monitors data, Power BI Eventstreams, and reports for specific patterns or thresholds (Iseminger et al., 2024). When predefined conditions are met, Data Activator triggers actions such as initiating Power Automate workflows or sending alerts. This capability enables you to create a digital nervous system that efficiently manages and tracks data. You can set up triggers for actions like Power Automate flows, Teams notifications, and emails without needing to rely on developers or IT teams (Iseminger et al., 2024). Data Activator enhances business agility and reduces operating costs.

Key Features of Data Activator:

- **Events:** Data Activator views data sources as a series of events reflecting the status of specific objects. These events can range from frequent occurrences (e.g., IoT sensor data) to rare ones (e.g., package scans in shipping).

- **Objects:** Objects are entities Data Activator interacts with, such as vehicles, parcels, or concepts like ads or sessions. By linking event streams to an object, you create a Reflex item, selecting an object ID column and defining the object's properties.

- **Triggers:** Triggers monitor events and data, activating predefined actions when certain conditions within the events are satisfied.

- **Properties:** Properties apply uniform logic across different triggers. For example, you could use a property on a smart lighting system to measure average brightness during peak activity periods, allowing multiple triggers to adjust lighting and conserve energy.

Chapter 7: OneLake & Lakehouse Unification

One of the most significant features of Microsoft Fabric is its seamless integration of OneLake and Lakehouse architecture. Microsoft Fabric Lake, known as OneLake, serves as the foundational layer for all essential services within the Fabric ecosystem. It acts as a centralized hub for storing all organizational data across various Fabric experiences, ensuring that users have a common platform to access and manage their data efficiently. Additionally, OneLake is built on Azure Data Lake Storage (ADLS) Gen2, a robust infrastructure that supports scalability, security, and high availability. This architecture allows Fabric to function as a unified SaaS product, simplifying user interactions and eliminating the need for deep understanding of complex underlying infrastructure concepts, such as redundancy, Azure Resource Manager, role-based access control (RBAC), and resource groups. By removing traditional data silos commonly found in modern applications, OneLake provides a cohesive storage solution that enhances data discovery and sharing. It also ensures consistent and centralized compliance with security policies and regulatory requirements.

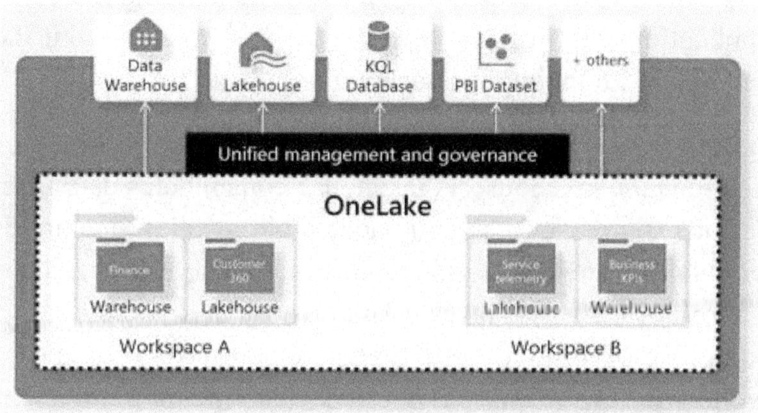

Figure 51 Onelake and Lakehouse Unification

OneLake utilizes a hierarchical management system to streamline data organization. The data lake is enhanced by Microsoft Fabric, which alleviates the need for pre-provisioning resources. Notably, Fabric operates with a single data lake per tenant, resulting in a unified file-system namespace that simplifies management across cloud resources, regions, and users. Within this hierarchy, the tenant is positioned at the top and serves as the root of OneLake, allowing users to create multiple workspaces as needed. This structure empowers users to efficiently load data into their Lakehouses, facilitating quick processing, analysis, and collaboration. All Fabric compute experiences are inherently connected to OneLake, meaning that individual functionalities-such as Data Engineering and Data Warehousing-utilize OneLake as their native storage without requiring additional configuration. Furthermore,

OneLake features a Shortcut functionality that enables users to easily mount existing PaaS storage accounts into OneLake. This feature simplifies data sharing between applications and users, allowing access to data without the necessity of copying or relocating it.

A core concept that underpins the diverse applications of Fabric is the lakehouse. A lakehouse consolidates data from various sources into a single accessible location, making it available to anyone within your organization who utilizes Azure-hosted data lakes. This architecture provides a comprehensive solution for managing large volumes of data while offering a unified interface that includes tools for storing, managing, and analyzing your data effectively.

Fabric's implementation of the lakehouse architecture supports Delta tables, which necessitates ensuring that any data within the lakehouse is formatted correctly. Once data is imported, users can leverage notebooks to explore and analyze the data, employing code to extract insights that can benefit other areas of the organization. Alternatively, users can access lakehouse data from other applications using a SQL endpoint, thereby enhancing interoperability. OneLake is compatible with tools such as Azure Databricks and Azure HDInsight, utilizing existing Gen 2 Azure Data Lake Storage APIs for enhanced functionality.

To create a lakehouse, users can either utilize the dashboard or an existing Fabric workspace. Data can be loaded using various methods tailored to the source type. The simplest approach is to upload data directly from a PC; however, the built-in copy tool is particularly advantageous, as it converts data into Delta tables. Users can also utilize Power BI's dataflow tool to connect and transform data from other platforms or employ Spark code for loading data into their Lakehouse's.

Microsoft Fabric empowers users to harness time-based, semi-structured data for real-time analytics, eliminating the need for separate tools for long-term and operational analysis. Users can interact with the same data in different ways to address varying analytical needs. Operational analytics facilitates rapid identification and resolution of issues, while the same datasets can also be leveraged for machine learning applications and report-based data analysis, integrating with additional data sources as required.

OneLake offers flexibility by not requiring all source data to be stored within it; users can create shortcuts to connect to external storage locations. Shortcuts function like symbolic links within the data lake, allowing users to work with data without necessitating its storage in Azure. This capability reduces the risks associated with data duplication and enables management of access to line-of-business systems directly through the Fabric dashboard. When a shortcut is created, it appears as a folder-specifically, a table folder for structured data and a file folder for unstructured data. If the shortcut points to Delta or Parquet formatted data, Fabric automatically recognizes it as a table, loading its metadata for seamless handling and management.

Many businesses are increasingly transitioning toward a single repository for all their data, and Microsoft is responding to this growing demand with Microsoft Fabric. By leveraging open standards such as Delta and Parquet, Microsoft has developed a solution that empowers organizations to create and operate data lakes using their existing data platform skills. This approach is designed to support both data warehouse analytics and machine learning initiatives, positioning businesses to harness the full potential of their data assets.

Chapter 8: Benefits of Microsoft Fabric

The various features and experiences of Microsoft Fabric highlight five key benefits:

Unified Analytics Platform

Microsoft Fabric stands out as a comprehensive, all-in-one analytics platform designed to streamline the complexity of managing multiple analytics subsystems. Traditionally, analytics projects involve a multitude of products from different vendors, creating challenges in integration, compatibility, and cost. This often results in a fragmented system that is error prone and inefficient. Fabric addresses these issues by providing a single, unified product with a cohesive architecture and user experience. This integration minimizes the hassle and cost associated with disparate systems, offering a seamless and consistent analytics experience. Furthermore, as a SaaS solution, Fabric automatically connects and enhances all its services and functions, allowing users to quickly set up a Fabric account and start deriving value from their data within minutes.

Microsoft Fabric's Open and Lake-Centric Approach

Traditional data lakes can be notoriously complex, presenting challenges in building, integrating, governing, and operating them effectively. They are often susceptible to vendor lock-in and data duplication due to the use of incompatible formats across different products within the data lake. Microsoft Fabric overcomes these hurdles with its OneLake architecture. OneLake serves as a centralized hub where all Fabric workloads seamlessly connect, ensuring optimal data organization. The data within OneLake is automatically indexed to facilitate compliance, management, sharing, and discovery. Fabric's use of open data formats across all tiers and workloads ensures that users only need to load their data into OneLake once, regardless of the format. This approach supports both structured and unstructured data, promoting efficiency and eliminating the complexities of traditional data lake management.

Microsoft Fabric's AI Integration

Microsoft Fabric incorporates advanced AI technology through the Azure Azure OpenAI service, enhancing how users interact with and utilize their data. AI capabilities are embedded throughout Fabric's layers, including the Copilot feature available in every experience. Copilot enables users to structure data pipelines, design dataflows, write code and functions, develop machine learning models, and generate analytical insights through natural language interactions. These AI services are designed with robust security measures, ensuring that user data remains protected. For example, Copilot's base language models are configured not to train on tenant data, thereby maintaining data privacy and security.

Fabric's Flexibility Across Enterprises

Microsoft Fabric is designed to be universally applicable, regardless of business type or size. It integrates seamlessly with all Microsoft 365 applications, ensuring broad usability across different organizational environments. For example, Power BI is effectively integrated with popular Microsoft applications such as PowerPoint, Teams, Excel, and SharePoint. This integration facilitates easy access to data from OneLake, making it convenient for users to leverage Microsoft 365 applications to gain insights. For example, Excel users can quickly search and analyze data with just a click, turning their everyday applications into powerful tools for data-driven decision-making.

Cost-Efficient Analytics through Unified Computing

Modern analytics projects often involve disparate products from various vendors, leading to inefficiencies due to separate computing capacities for different systems like business intelligence (BI), data warehousing, and data integration. This segregation results in wasted resources, as idle capacity in one system cannot be utilized by another. Microsoft Fabric addresses this issue by providing a unified computing pool that supports all workloads, thereby optimizing resource use. Users can create comprehensive analytics solutions that leverage any workload without conflicts, leading to significant cost savings. By avoiding idle compute capacities and consolidating resources, Fabric enables more efficient and economical analytics operations.

Chapter 9:
The Power of Integration

Microsoft Fabric and Data Teaming

Microsoft Fabric serves as a unified management and governance solution that simplifies the complexities of data projects for professionals. By eliminating data silos and reducing the need for multiple systems, Fabric enhances collaboration among data professionals, fostering a more cohesive and efficient workflow.

Historically, data engineers and data analysts had distinct roles, often requiring extensive communication to ensure the data engineer's semantic models were accurately designed to support the data analyst's need for meaningful business insights. This separation could lead to inefficiencies and misalignment between teams.

With Microsoft Fabric, data professionals can now collaborate within a single SaaS product, bridging the gap between different roles and aligning their efforts with business objectives. Data analysts benefit from easier data transformations upstream, facilitated by the integration with Data Factory, allowing them to access and manipulate data more efficiently.

For data engineers seeking to simplify their semantic model curation or expand their expertise into data science, Fabric provides a robust solution tailored to meet organizational requirements. Data engineers can leverage Fabric's features to enhance their workflows and broaden their skill sets.

Data analysts, who previously spent considerable time on data transformations before creating Power BI reports, can now connect with data more seamlessly and trace data lineage with DirectLake mode. This streamlines their process and improves overall efficiency.

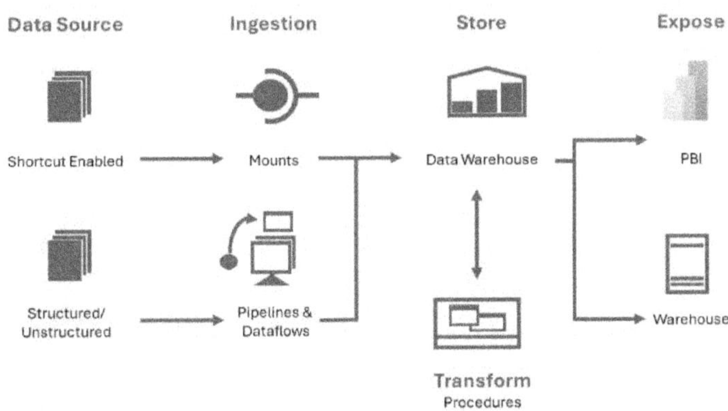

Figure 52 Data Sources Exposed for Ease of Use

Data scientists can leverage Power BI's interactive reporting capabilities, integrating native data science methods to deliver insights more effectively. The advanced features in Fabric enable a more intuitive approach to data science, enhancing the analytical capabilities available to data scientists.

Fabric's SaaS platform offers the flexibility to run any type of workload or job instantly and with minimal planning. This adaptability allows organizations to adjust resources as needed, making it easier to respond to evolving business demands.

Furthermore, Fabric embraces the low-to-no-code philosophy that has driven success on the Power Platform, applying it to its SaaS services. This approach maintains scalability and quality across data science, data warehousing, data ingestion and preparation, and analytics, while also providing visual coding options that can help users overcome previous barriers to advancement.

Chapter 10: Copilot in Microsoft Fabric

Copilot and Generative AI Features

Copilot, along with other preview features utilizing generative AI, can significantly enhance how you interact with data, uncover insights, and create visuals and reports in Microsoft Fabric and Power BI. To access these features, you must first enable Copilot within your Microsoft Fabric environment.

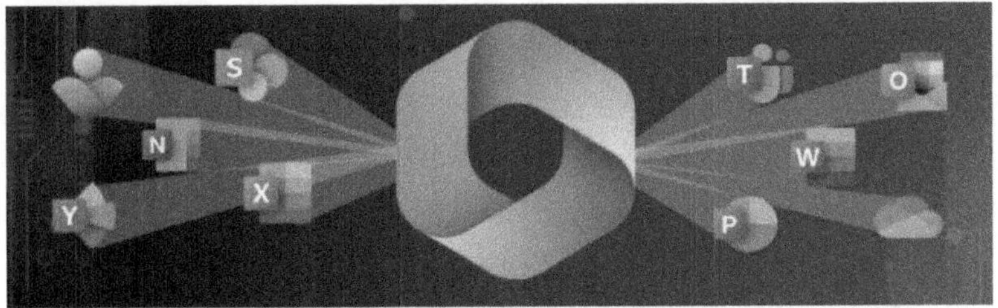

Figure 53 Microsoft Co-Pilot

Copilot for Data Engineering and Data Science

Copilot for Data Engineering and Data Science offers an AI-powered toolset designed to streamline workflows for data professionals. It facilitates faster code completion, automates routine tasks, and provides code templates that adhere to industry standards, aiding in the development of robust data pipelines and complex analytical models. By leveraging advanced machine learning algorithms, Copilot delivers contextual code suggestions tailored to the specific task at hand, enhancing coding efficiency and ease. Whether you are involved in data preparation or generating insights, Microsoft Fabric Copilot acts as an intelligent assistant, reducing the workload on engineers and scientists and accelerating the transition from raw data to actionable conclusions.

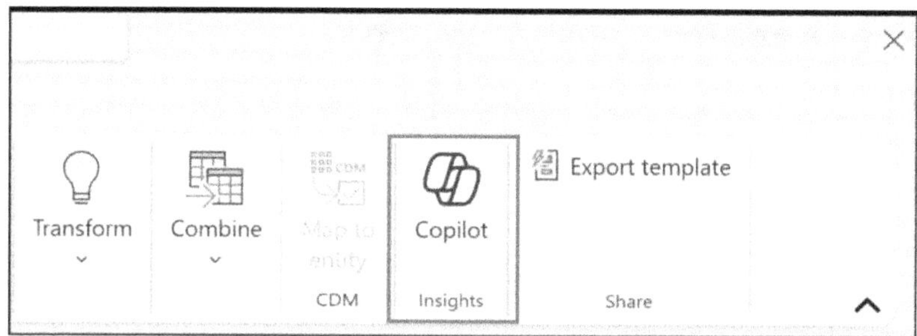

Figure 54 Co-Pilot for Data Factory

Copilot for Data Factory

Copilot for Data Factory is an AI-driven toolset that helps data wranglers of all experience levels optimize their workflows. It generates intelligent code to facilitate data transformation and provides explanations for complex tasks, making it easier to understand and execute data operations.

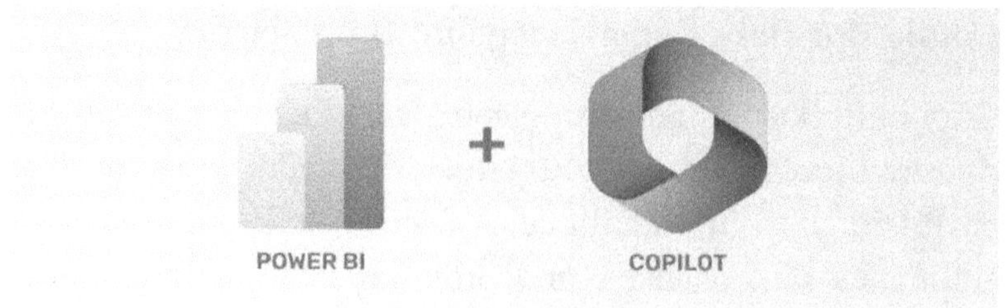

Figure 55 Co-Pilot and Power BI

Copilot for Power BI

With the new generative AI features in Power BI, you can create reports by selecting or suggesting topics to Copilot. Additionally, Copilot assists in summarizing report pages and finding synonyms to enhance question-and-answer functionalities, improving the clarity and effectiveness of your reports

Copilot in Power BI allows business managers to generate reports and obtain insights through natural language inputs. By specifying the needed visuals, insights, and calculations, Copilot can automatically create the report, analyze the data, and offer interactive visualizations.

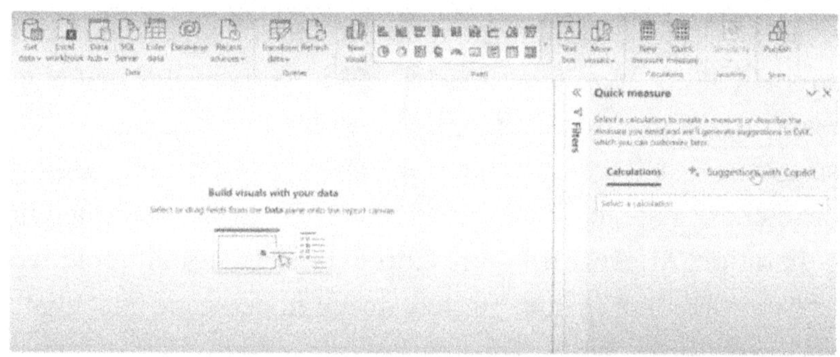

Figure 56 Co-Pilot in Fabric

Copilot in Fabric

Microsoft Fabric, along with Power BI, introduces new possibilities for data transformation, analysis, and visualization through Copilot and other generative AI features available in preview. These tools revolutionize how data is handled, offering advanced capabilities for generating insights and creating reports.

Enabling Copilot

To utilize Copilot features within Microsoft Fabric for your business, you need to enable Copilot first. This activation will allow you to leverage the full range of Copilot's capabilities and integrate AI-enhanced functionalities into your data workflows.

Copilot for Data Science and Data Engineering

Copilot for Data Engineering and Data Science provides an AI-enhanced toolset specifically designed for data professionals. It supports intelligent code completion, automates repetitive tasks, and offers industry-standard code templates for building reliable data pipelines and sophisticated analytical models. With advanced machine learning algorithms, Copilot offers contextual code suggestions tailored to your current task, improving coding efficiency and ease. From data preparation to insight generation, Microsoft Fabric Copilot serves as an interactive assistant, reducing the burden on engineers and scientists and expediting the transformation of raw data into valuable insights.

Chapter 11:
AI Overview

Computational Intelligence Approaches

Artificial Intelligence (AI) has undergone a remarkable evolution in recent years, ushering in an era where machines not only learn but also make informed decisions autonomously. At its core, AI refers to the capacity of computers to perform tasks that are typically associated with human intelligence. This includes recognizing objects, making decisions, translating languages, and more. Initially, AI research focused on designing systems capable of computing, learning, and solving complex problems through logical reasoning.

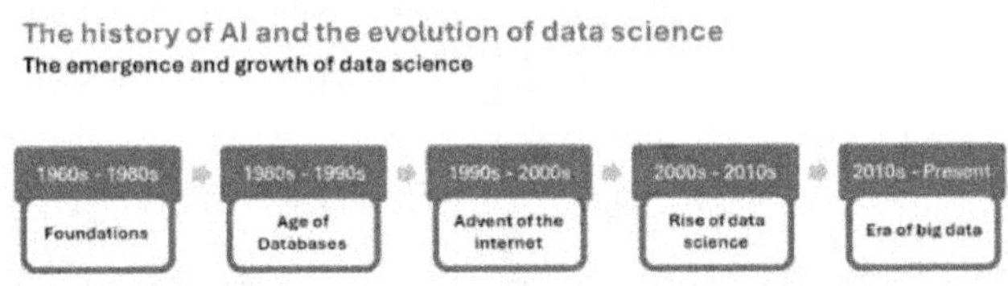

Figure 57 History of Data Science Evolution

In the modern landscape, the significant advancements in AI can be largely attributed to breakthroughs in machine learning (ML) and deep learning (DL). These data-driven methodologies allow computers to learn automatically from vast datasets, extracting patterns and insights without the need for explicit programming. While AI encompasses a broad spectrum of technologies aimed at achieving human-like intelligence, ML and DL are pivotal subfields that harness data and algorithms to empower machines to learn and evolve independently. This evolution contributes significantly to the rapid advancements observed in the AI domain.

Artificial Intelligence: Definition and Development

Artificial Intelligence refers to the ability of computer systems and machines to perform tasks that are generally thought to require human intelligence. This involves cognitive abilities such as visual perception, problem-solving, decision-making, and language translation. AI is a field within computer science focused on creating machines and systems that can perform cognitive functions at or above the level of human performance. AI systems aim to replicate human intelligence but lack true consciousness, self-awareness, and general understanding. They are programmed for specific tasks within set limits. For example, an AI trained for speech recognition would need extensive retraining to play chess. Moreover, AI differs from natural intelligence that has evolved over millions of years in humans and animals.

The foundational concepts and methodologies of artificial intelligence date back to the 1950s, when researchers began exploring whether machines could simulate the human brain's capabilities. According to Radanliev (2024), early milestones included the development of logical theories and problem-solving algorithms by pioneering scientists such as

John McCarthy and Allen Newell. In the 1980s, AI research yielded some expert systems, particularly in specialized domains such as medical diagnosis; however, progress was often hindered by limited computing power.

The modern era of AI commenced in the late 2000s, fueled by exponential increases in processing power and advancements in Graphics Processing Units (GPUs). The availability of massive datasets further facilitated the training of machine learning algorithms, enabling them to perform at unprecedented levels (Huawei Technologies Co., Ltd, 2022). A pivotal moment occurred in 2015 when deep learning systems began to surpass human-level performance in image recognition tasks. Since then, AI has rapidly advanced across various fields, including computer vision, speech recognition, machine translation, and strategic game-playing. Today, AI is seamlessly integrated into many technologies and continues to be applied in new and innovative domains.

Approaches to Achieving Artificial Intelligence

Two primary approaches have emerged in the quest to achieve artificial intelligence: symbolic AI and connectionist AI. Symbolic AI, often referred to as logic-based AI, utilizes logical rules and symbols to represent knowledge and execute tasks through logical reasoning and expert systems. An early example of symbolic AI is MYCIN, an expert system developed in the 1970s to diagnose infectious diseases and recommend treatments (Press, 2023). While symbolic AI has demonstrated success in narrow, specialized domains, it struggles with tasks that require common sense reasoning or perceptual understanding.

In contrast, connectionist AI, also known as computational intelligence, draws inspiration from the workings of the human brain and nervous system. This approach employs large networks of simple processing units, such as artificial neural networks, to uncover complex patterns in data through connection strengths known as weights (Shao & Shen, 2023). Connectionist AI has shown remarkable success in pattern recognition tasks, including computer vision, natural language processing, and predictive analytics. By mimicking the interconnected nature of neurons in the brain, connectionist models enable machines to learn from data in ways that resemble human cognitive processes.

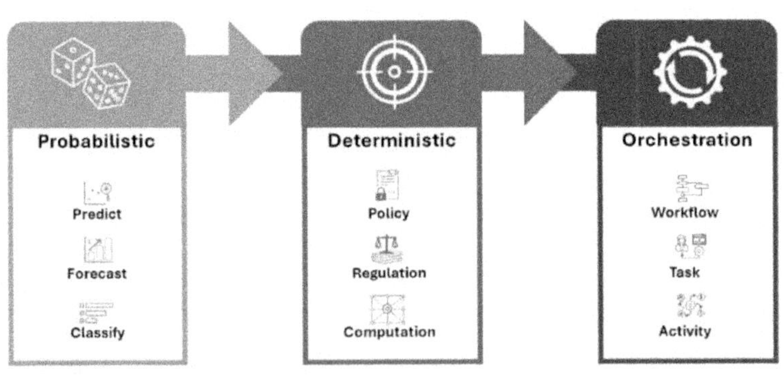

Figure 58 AI Approaches

Modern AI and General vs. Narrow AI

Modern artificial intelligence (AI) has achieved remarkable advancements in several key areas, including deep learning, conversational systems, and probabilistic and qualitative reasoning. In the field of computer vision, AI now excels at solving complex problems in image recognition, classification, and perception (Huawei Technologies Co., Ltd., 2022). Contemporary AI systems can analyze photos and videos to identify objects, faces, and scenes with accuracy that rivals or exceeds human performance.

Natural Language Processing (NLP) is another area where AI has made significant strides. NLP enables machines to understand, disaggregate, and organize human languages, from the molecular level of syntax and semantics to higher-level applications such as machine translation, chatbots, and text analysis tools. These advancements allow AI to engage in sophisticated interactions with users and process language data effectively.

The ultimate goal of AI research is to develop systems that exhibit general intelligence akin to human cognition, across a broad range of intellectual tasks. This ambitious pursuit, often referred to as General AI or Artificial General Intelligence (AGI), aims to create machines with the flexibility and adaptive capabilities of human intelligence. In contrast, Narrow AI (or Weak AI) focuses on specific tasks and problem domains, such as image recognition or language translation, without pursuing the broader problem-solving abilities characteristic of human cognition (Radanliev, 2024). While Narrow AI excels at specific applications, the pursuit of General AI involves exploring advanced methodologies, such as whole-brain emulation and developmental learning models, to recreate comprehensive human-like intelligence.

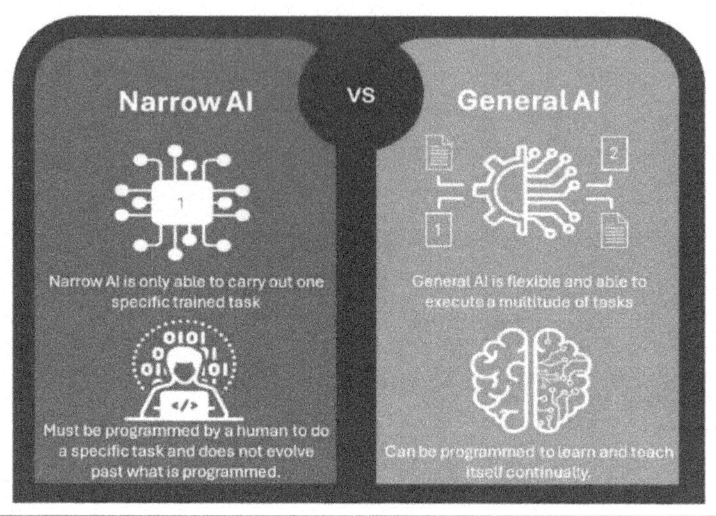

Figure 59 Narrow vs. General AI

Machine Learning

Definition of Machine Learning

Machine learning is a pivotal branch of AI that involves enabling systems to learn and make predictions from data without being explicitly programmed for each task. The essence of machine learning lies in developing algorithms that process input data and generate predictions through statistical analysis, rather than relying on predefined, task-specific instructions (Taye, 2023). Machine learning algorithms use "training data," or sample datasets, to build mathematical models that make predictions or decisions on new, previously unseen examples. These algorithms identify patterns within large datasets, allowing them to generate predictions and improve their performance over time. Unlike traditional programming, where engineers manually craft rules, machine learning algorithms refine their capabilities automatically through experience, adapting to new data without human intervention.

Core Concepts of Machine Learning Algorithms

Machine learning algorithms are fundamentally mathematical or computational models that are optimized iteratively using training data. According to Taye (2021), these algorithms aim to approximate the relationships between input and output variables to make accurate predictions on new examples. Models are often initialized with random parameters and refined through iterative processes, such as multiple passes of the training data. Techniques such as gradient descent adjust model parameters, including neural network weights, to minimize errors and improve accuracy with each iteration. Once trained, these models can be applied to new data to make predictions or informed judgments. Common algorithmic approaches include decision trees, support vector machines, regression analysis, clustering, dimensionality reduction, neural networks, and ensemble methods that combine the outputs of multiple models to enhance performance (Sarker, 2021).

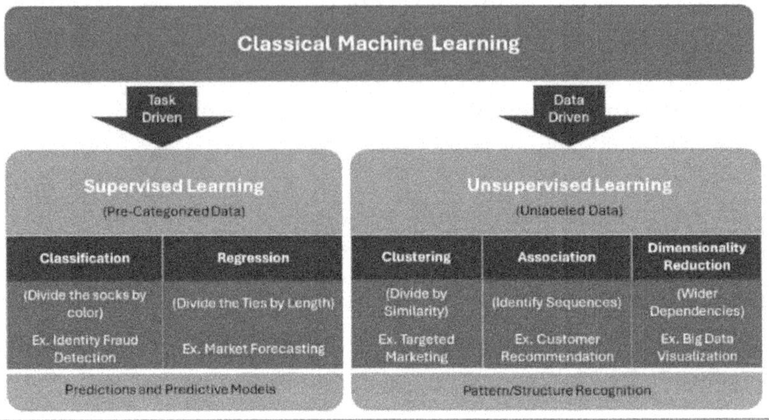

Figure 60 Classic Machine Learning

Types of Machine Learning Algorithms

Supervised Machine Learning

In supervised machine learning, the algorithm learns from labeled training data, where each input instance is paired with a known output. The goal is to identify patterns and relationships between the input data and the corresponding outputs to build a mathematical model (Pugliese et al., 2021). Once trained, this model can predict outcomes for new, unlabeled data based on the patterns it has learned. For example, in classification tasks, algorithms like logistic regression are used to categorize data into predefined classes. A practical example is a supervised machine learning system trained on emails labeled as "spam" or "not spam." By analyzing the content and metadata of numerous labeled emails, the algorithm learns to recognize patterns that distinguish spam from non-spam emails, enabling it to accurately classify new emails using these learned patterns (Sarker, 2021b).

Unsupervised Learning

Unsupervised learning is a machine learning approach that seeks to uncover hidden patterns or structures within unlabeled input data. Unlike supervised learning, where models are trained on labeled datasets with predefined categories, unsupervised learning algorithms autonomously determine the underlying structure of the data without explicit human guidance (Sarker, 2021a). This approach is particularly useful in scenarios where labeled data is scarce or unavailable. Common applications include market segmentation, where customers are grouped based on spending patterns without predefined classes; document classification by topic based on word-frequency analysis; and anomaly detection to identify unusual patterns indicative of fraud.

Unsupervised learning techniques encompass various methods, such as clustering algorithms like k-means clustering, which identify natural groupings within data by partitioning it into clusters based on similarity measures. Dimensionality reduction techniques, such as Principal Component Analysis (PCA), are used to reduce the number of features while preserving essential information, making data easier to visualize and analyze. Anomaly detection algorithms identify data points that significantly deviate from the norm, which is crucial for applications such as fraud detection and network security.

Semi-Supervised Machine Learning

Semi-supervised learning combines both labeled and unlabeled data to enhance model training, particularly when fully labeled datasets are limited in size. This approach leverages a small amount of labeled data alongside a larger pool of unlabeled data to build more robust models (Pugliese et al., 2021). Semi-supervised learning methods often employ self-training algorithms that initially classify unlabeled data using a small labeled dataset. The most confident predictions are then added to the labeled set, and the model is retrained iteratively to refine its performance.

A practical example of semi-supervised learning is analyzing online behavior data for spam detection. Given that only a small fraction of the data is labeled, semi-supervised learning algorithms propagate labels from labeled data to large amounts of unlabeled data, improving the model's accuracy in identifying spam. This approach effectively leverages the abundance of unlabeled data to enhance learning and improve predictive performance.

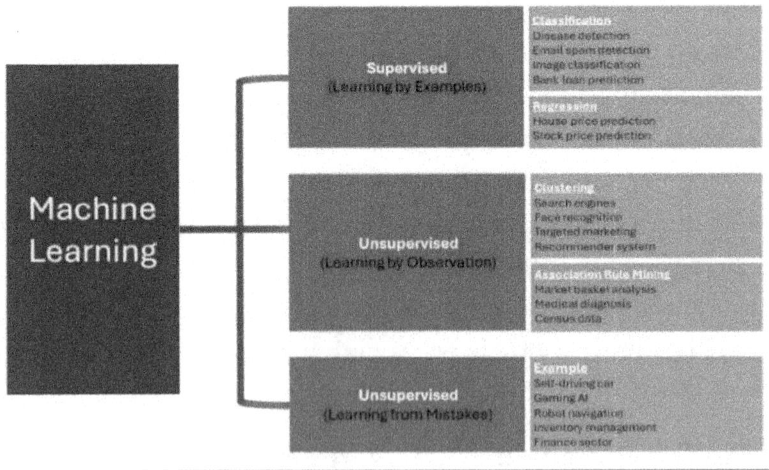

Figure 61 Types of Machine Learning

Reinforcement Machine Learning

Reinforcement learning focuses on goal-oriented learning in interactive environments, where an agent learns through trial and error. The agent performs actions, receives feedback in the form of rewards or penalties, and adjusts its behavior to optimize performance in complex, dynamic settings (Sarker, 2021a). Unlike supervised learning, which relies on explicit examples, reinforcement learning uses exploratory trials to discover the best strategies or policies.

The key components of reinforcement learning include defining policies that map situations to optimal actions and using feedback to improve these policies over time. Applications of reinforcement learning span various domains, including game playing, where agents learn to outperform human players in complex video games; robot control for autonomous navigation and manipulation tasks; dialogue generation for natural language interactions; and sequencing tasks for optimal decision-making. For example, deep reinforcement learning techniques have enabled agents to achieve super-human performance in 3D video games by learning from experience without relying on pre-designed levels or telemetry data.

Definition and Characteristics of Deep Learning

Deep learning is a subset of machine learning that employs deep neural networks, which consist of multiple processing layers. Inspired by the structure and function of the human brain, deep learning algorithms process large volumes of raw data through hierarchical layers of non-linear transformations to learn and extract complex features (Sarker, 2021b). This hierarchical learning approach allows deep learning models to capture intricate patterns and representations from data progressively.

Deep learning distinguishes itself by using neural network architectures with numerous hidden layers between the input and output layers. These architectures enable the learning of increasingly abstract features from data samples such as images, videos, audio, text, or scientific data. According to Alzubaidi et al. (2021), deep learning models undergo a training process called backpropagation, where lower layers initially detect basic attributes, and higher layers integrate these attributes into more complex concepts. This hierarchical process results in a model capable of detecting simple features and forming complex conceptual representations. Deep learning has demonstrated exceptional performance in domains such as bioinformatics and computational physics, where it automatically identifies intricate patterns in large datasets.

Major Deep Learning Architectures

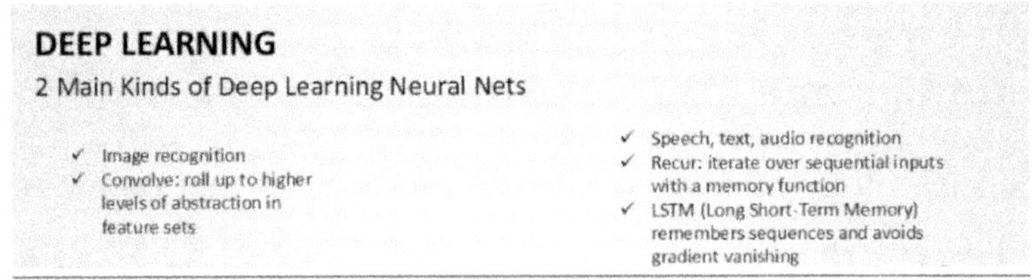

Convolutional Neural Networks (CNNs)

Convolutional Neural Networks (CNNs) are particularly effective for processing image and visual data due to their ability to learn spatial hierarchies of patterns. CNNs utilize convolutional layers to perform mathematical operations called convolutions, which extract features from local regions of input images (Sarker, 2021b). These features are then simplified by max-pooling layers while retaining essential spatial information. By stacking multiple convolutional and pooling layers, CNNs can identify increasingly complex patterns and structures within images.

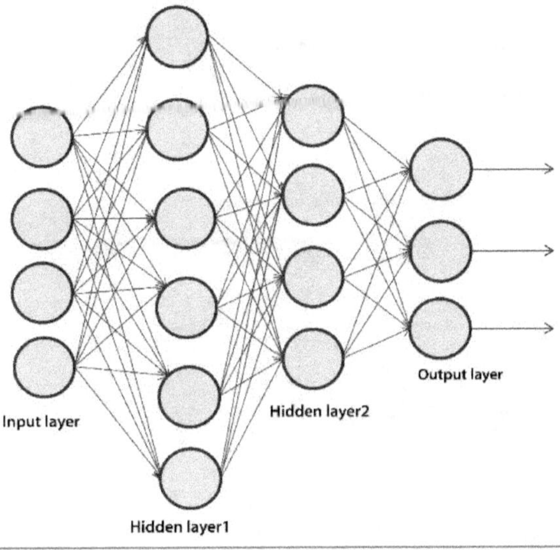

Figure 62 Convolutional Neural Networks

Notable CNN architectures, such as AlexNet, ResNet, and VGG, have achieved remarkable performance on large-scale visual recognition tasks, including ImageNet classification, surpassing human-level accuracy (Alzubaidi et al., 2021). CNNs are widely used in various visual applications, including image categorization, object detection within scenes, facial recognition, and medical image analysis. Their ability to learn and generalize from vast amounts of visual data has made them a cornerstone of modern computer vision.

Recurrent Neural Networks (RNNs)

Recurrent Neural Networks (RNNs) are particularly well-suited for processing sequential data, such as text, speech, audio, time series, and genomic sequences, due to their internal feedback loops. These loops enable RNNs to analyze inputs over time and exhibit dynamic, temporal behavior that distinguishes them from traditional feedforward neural networks.

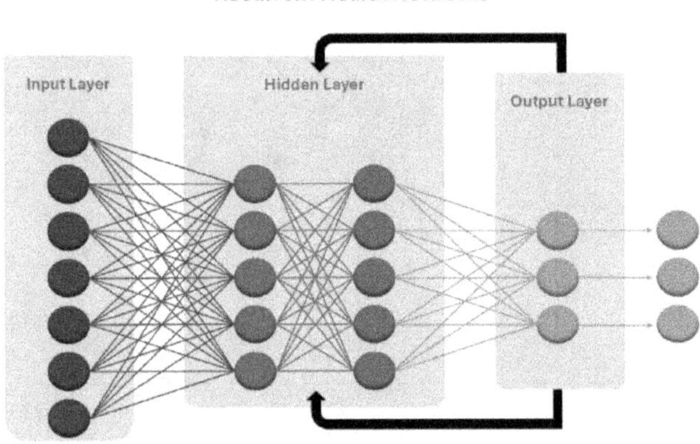

Figure 63 Recurrent Neural Network

The ability of RNNs to maintain an internal memory state enables them to process variable-length sequences, making them adept at tasks where context and order matter. However, standard RNNs face challenges in learning dependencies among inputs that are distant within a sequence, a problem known as the vanishing gradient problem. More advanced RNN architectures, such as Long Short-Term Memory (LSTM) networks and Gated Recurrent

131

Units (GRUs), address this limitation by incorporating gating mechanisms that regulate the flow of information within the network. These mechanisms help the network retain important information over long sequences and discard irrelevant data (Sarker, 2021b). Consequently, RNNs and their advanced variants have found widespread applications in fields involving sequential data, including speech recognition, handwriting synthesis, and forecasting.

Training Deep Learning Systems via Big Datasets

Deep learning models are characterized by their extensive use of neural network weights, which can number in the millions or even billions. Training these models involves iteratively adjusting these weights to minimize a loss function that quantifies the error between the model's predictions and the actual outcomes (Sarker, 2021b). According to Alzubaidi et al. (2021), this process is achieved using large labeled datasets and a technique known as backpropagation. Backpropagation calculates the gradients of the loss function with respect to the model's weights, enabling efficient updates to minimize the loss. Models are initialized with random weights and then progressively improved over multiple "epochs," in which the model passes over the training data repeatedly. For example, image recognition networks may be trained on datasets containing millions of images, each labeled with thousands of categories. During training, batches of images are fed into the network, and backpropagation uses classification errors to adjust the weights, thereby enhancing the network's ability to classify images accurately. This iterative process demands substantial computational resources, such as GPU clusters, and can take days or even weeks to converge. Once trained, the network can generalize its learned patterns to recognize novel images based on the specific visual features it has identified during training.

Relationships Between AI, ML, and Deep Learning

Hierarchical Relationship

The fields of artificial intelligence (AI), machine learning (ML), and deep learning (DL) are organized hierarchically by breadth and level of abstraction.

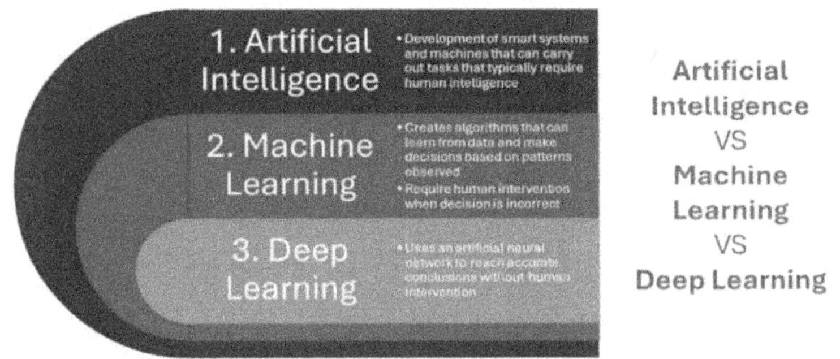

Figure 64 Relationship between AI, ML, and DL

AI is the most encompassing concept, aiming to develop systems capable of tasks requiring human-like intelligence, such as reasoning, problem-solving, and other cognitive functions (IBM, 2023). Within this broad objective, machine learning and deep learning represent more specialized methodologies for achieving intelligent behavior. Machine learning focuses on the development and application of algorithms that enable systems to learn from data and make predictions, without relying solely on explicitly programmed instructions. Deep learning, a subset of machine learning, utilizes neural network architectures with multiple layers to learn hierarchical representations from raw data.

This hierarchical structure places AI at the top level, with machine learning and deep learning as technical subsets within the broader AI paradigm (IBM, 2023). While AI research encompasses a wide range of approaches to creating intelligent systems, machine learning and deep learning specifically leverage data and algorithms to drive learning and decision-making processes. These approaches enable intelligent behaviors to emerge from experience, reducing the need for manual programming of rules and behaviors. Consequently, modern AI methods have accelerated advancements by focusing on self-supervised learning and automated pattern recognition, contributing to the ongoing pursuit of achieving human-level intelligence.

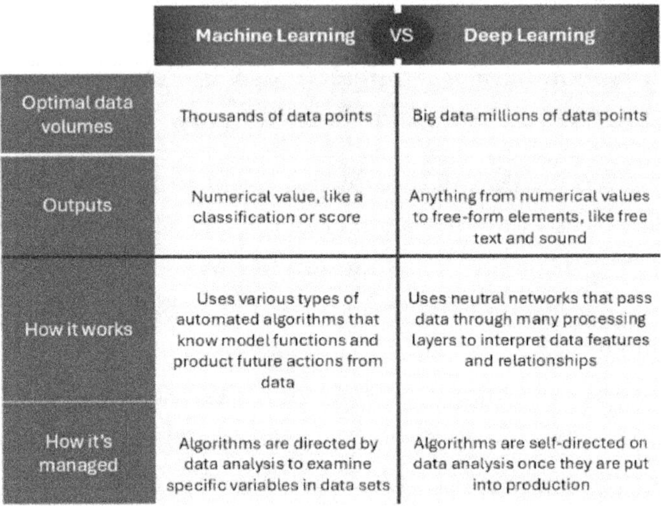

Figure 65 Machine Learning vs Deep Learning

AI, Machine Learning, and Deep Learning

Machine learning and deep learning are both subsets of artificial intelligence that employ specific algorithmic techniques and datasets to enable intelligent task performance (Jagdale et al., 2022). Unlike traditional AI approaches that rely heavily on human-defined rules and logic, machine learning and deep learning leverage data-driven methods to learn and make predictions. Machine learning algorithms learn from training examples provided, with varying degrees of human involvement in data pre-processing, feature engineering, and model tuning (IBM, 2023). Deep learning takes this a step further by enabling automatic representation learning from raw inputs through layers of abstraction, thereby minimizing the need for manual feature design.

The need for and amount of human involvement in machine learning and deep learning systems have evolved. Early supervised learning models required extensive human labeling of data. However, advancements have led to methods that increasingly utilize unsupervised, self-supervised, and reinforcement learning techniques to discover patterns without complete labeling. This shift reduces the need for precise human engineering to define target variables or rules. Moreover, modern deep learning models, particularly those used for generation or reinforcement tasks, can exhibit unexpected behaviors that their designers did not anticipate. These models demonstrate a high degree of autonomy, functioning effectively based on the training data and network architecture defined during their development (Jagdale et al., 2022).

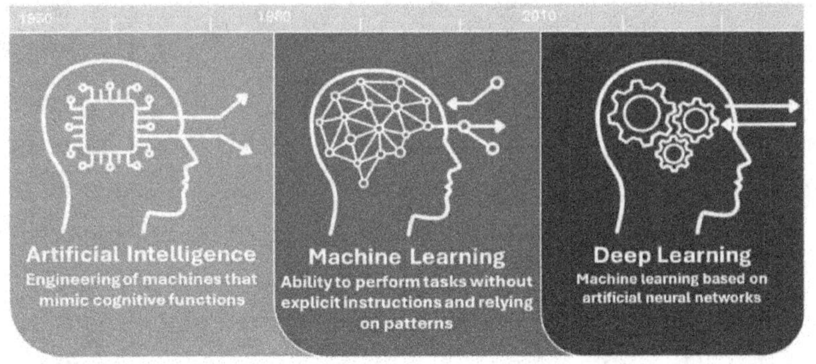

Figure 66 AI -> ML -> DL

DL Expanding Upon Machine Learning

Deep learning represents a significant advancement over traditional machine learning by employing neural network architectures to learn features directly from raw data. According to Alzubaidi et al. (2021), rather than relying on human engineers to manually specify relevant features for modeling, deep neural networks are designed to autonomously discover intricate patterns and representations in data through their multilayered structure. These networks operate hierarchically: the initial layers identify basic, lower-level features, such as edges in image data, while successive layers progressively build more complex representations, culminating in high-level abstractions. This hierarchical learning mimics the human brain's processing approach, allowing deep learning models to capture and model highly complex nonlinear relationships in data. As a result, deep learning is particularly effective at handling large volumes of unstructured or semi-structured datasets, which traditional machine learning methods often struggled to process effectively.

The ability of deep neural networks to automatically learn and extract meaningful features from raw data has revolutionized various applications. For example, tasks such as speech recognition, object detection, and machine translation, which were previously beyond the scope of classical machine learning techniques, have become feasible and highly effective thanks to deep learning's capacity to generate informative intermediate embeddings (Sarker, 2021a). This advancement has liberated algorithms from the constraints of manual feature engineering, leading to significant progress in both research and commercial applications of artificial intelligence. Deep learning has thus emerged as the dominant approach in machine learning, driving innovations across natural language understanding, computer vision, and time-series analysis.

Interchangeable Use of Terms

Artificial intelligence (AI), machine learning (ML), and deep learning (DL) are interconnected fields that are often used interchangeably in both industry and research contexts. This interchangeable use of terms can be attributed to the rapid evolution of these fields and the way deep learning has built upon and enhanced traditional machine learning techniques (IBM, 2023). Additionally, there is a lack of standardized definitions and terminology across different organizations and individuals involved in these rapidly advancing areas. Consequently, AI is frequently used as an umbrella term that encompasses both machine learning and deep learning applications (Jagdale et al., 2022). Furthermore, deep learning is sometimes mistakenly equated with machine learning as a whole, even though it is technically a subset of machine learning techniques.

This overlap in terminology can create confusion for newcomers trying to grasp the relationships and distinctions between these domains. However, it also reflects their close connections: deep learning is a specialized subset within the broader fields of machine learning and AI. In practical industry applications, where the focus is often on solving real-world problems and creating innovative solutions, the exact distinctions between methodologies may be less critical than the ability to develop effective and scalable systems. Nonetheless, understanding the conceptual differences among AI, machine learning, and deep learning is essential for researchers and practitioners who aim to advance these fields, communicate new developments, or address emerging challenges. Recognizing these distinctions helps clarify the scope and potential of each approach in building more human-like intelligent systems.

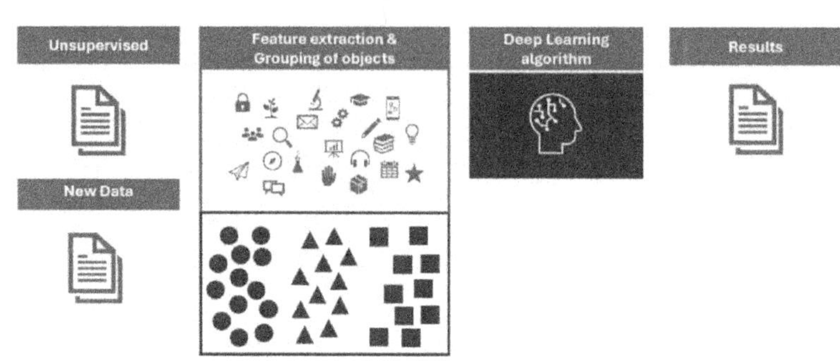

Figure 67 Deep Learning Process

ML and DL Techniques Enable Modern Apps

Machine learning and deep learning techniques have been pivotal in the development of many contemporary AI applications. By leveraging algorithms that learn from data rather than relying solely on explicit programming, these methods have brought significant improvements across various domains. For example, advancements in computer vision, natural language processing, diagnostics, and predictive analytics have been largely driven by machine learning's ability to recognize patterns, make inferences, and respond to queries (IBM, 2023; Jagdale et al., 2022). Deep learning, in particular, has accelerated progress by enabling feature learning directly from raw inputs, such as images, videos, text, and sensor data. This capability has enabled advances in areas such as autonomous vehicles, medical imaging, and online advertising.

The emphasis on self-supervised learning in machine learning and deep learning also allows these AI systems to continually improve as more data becomes available, unlike traditional rule-based programming approaches. This dynamic learning capability has led to applications becoming increasingly useful and robust over time.

For example, translation and recommendation systems can adapt and refine their outputs based on live user interactions without requiring extensive re-engineering (Jagdale et al., 2022). By reducing the need for manual programming, machine learning and deep learning have become fundamental enablers of AI's broad practical applications across business, science, and everyday life. Their continued evolution is crucial for driving future innovations in intelligent technologies.

In summary, artificial intelligence (AI), machine learning (ML), and deep learning (DL) are closely interrelated fields driving remarkable advances in intelligent systems. AI encompasses the overarching goal of creating computer systems with human-like intelligence, including reasoning, problem-solving, and other cognitive functions. Machine learning and deep learning offer specific methodologies within this framework, utilizing algorithms and neural networks to learn from data. While machine learning provided the initial breakthroughs, deep learning has further expanded learning capabilities through complex, layered neural networks. As data volumes continue to grow exponentially and computational power advances, deep learning, in particular, will remain a critical technique in the quest to achieve more human-like artificial intelligence.

Understanding the distinctions and relationships among AI, machine learning, and deep learning enhances our appreciation of their capabilities and limitations, guiding the responsible development and application of these technologies across sectors of society.

Chapter 12: Deployment Leveraging CI/CD on Azure

Overview of CI/CD Leveraging Azure

Continuous advancements in technology are transforming the design and development of automation tools, reducing the workload on software engineers and developers. Automation enhances the efficiency of the software development process by optimizing resource allocation, creative thinking, and effort. By adopting partial or full-scale automation, development teams can minimize the time spent on repetitive tasks, thereby reducing operational costs. The Continuous Integration and Continuous Deployment (CI/CD) pipeline is instrumental in realizing these benefits, delivering updated software versions while ensuring code quality. CI/CD processes significantly simplify delivery, reduce human error, and increase development efficiency by enabling continuous updates and maintenance through shorter feedback cycles. These iterations result in consistent, manageable modifications, minimizing disruptions during the deployment phases.

Incorporating CI/CD pipelines within Microsoft Azure technologies allows for automated deployment, artifact publishing, production deployment, approval gates, code commits, enhanced testing, and automated builds. This integration leads to faster feedback loops and more reliable, continuous code deployment.

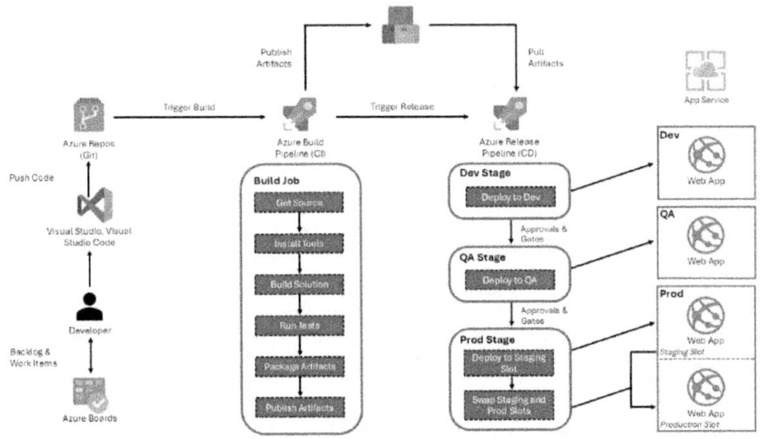

Figure 68 Continuous Integration & Continuous Deployment

What is CI/CD?

CI/CD stands for Continuous Integration and Continuous Deployment, a DevOps practice that automates the software delivery process. Continuous Integration (CI) involves frequent updates to a central branch, where developers add the latest working copies of the software at regular intervals. Build servers support CI by automatically verifying each merge, running unit tests, and reporting the findings to developers. CI gained significant traction in the late 1990s and became widely recognized in the early 2000s through Martin Fowler's advocacy. It is a best practice in agile software development, enhancing software quality by addressing implementation issues among collaborative developers.

Continuous Deployment (CD) extends CI by automatically deploying software to client environments after each change. CD employs a push-based approach with frequent and automated releases, ensuring that every update is instantly deployed to production environments without manual intervention. Together, CI/CD pipelines offer substantial benefits to the software engineering industry, including faster defect discovery, accelerated releases, and improved developer productivity.

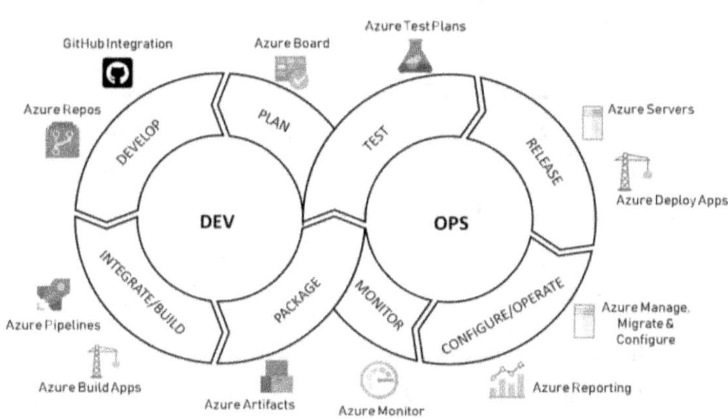

Figure 69 Azure Dev/Op Universe

Microsoft Technologies for CI/CD

Azure DevOps is a leading platform for automating software building and deployment processes. It offers a range of free and paid services tailored to various organizational needs. Azure DevOps is a comprehensive set of development tools that can build, test, and deploy solutions in any language,

whether on-premises or in the cloud. The key components of
Azure DevOps include:

1. **Azure Repos** provides cloud-hosted private Git
    repositories and version control tools to manage
    code. It facilitates code management through code
    reviews, pull requests, and branching. Developers
    can review and validate code modifications before
    deployment, ensuring that all changes pass the
    necessary tests. Azure Repos also supports the
    centralized version control system "Team
    Foundation Version Control" (TFVC), where
    historical data is stored on the server, creating path-
    based branches.

2. **Azure Pipelines** Azure Pipelines enable developers
    to build, test, package, and release infrastructure
    code and applications. They include three types of
    pipelines: Pull Request (PR) pipelines, CI pipelines,
    and CD pipelines. PR pipelines validate code, while
    CI pipelines add implementation testing and publish
    artifacts upon successful validation. CD pipelines
    deploy the build artifacts, run acceptance tests, and
    release the software to production environments.
    Azure Pipelines support multiple platforms and
    languages, and YAML is used to define pipelines,
    enhancing collaboration and streamlining the
    review process.

3. **Azure Artifacts** Azure Artifacts allow developers to
    manage and share various software packages, such
    as NuGet, npm, and Maven. They handle the entire
    package lifecycle, ensuring that the development

team uses the most secure and up-to-date package versions. Integrating CI/CD pipelines with package management enables automatic package absorption and publication, ensuring efficient dependency management.

4. **Azure Test Plans** Azure Test Plans provide tools for managing testing efforts throughout the software development lifecycle. They support both automated and manual testing, offering valuable insights into test results. Integration with CI/CD pipelines allows developers to automate test case execution during the build and release process, making testing an integral part of development.

5. **Azure Boards** Azure Boards are customizable and interactive tools for managing development projects. They support Agile, Kanban, and Scrum processes, and offer integrated reporting and configurable dashboards. Azure Boards' scalability makes them ideal for dynamic business environments. Integration with the CI/CD pipeline facilitates work tracking and ensures end-to-end traceability by linking work items with code commits, releases, and builds.

GitHub Actions: Enhancing Workflow Automation and Scalability

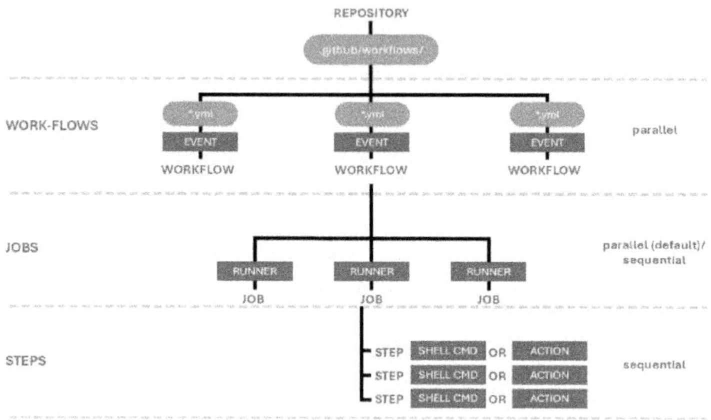

Figure 70 GitHub Actions

1. Workflow Automation GitHub Actions provides a robust platform for automating workflows, offering flexible and intuitive CI/CD features. These capabilities allow developers to seamlessly integrate Continuous Integration (CI) into their GitHub flow using custom templates or by creating their own CI/CD workflows. With native GitHub capabilities, developers can access thousands of pre-built and tested automations available in the GitHub Marketplace or write custom code using YAML files. YAML files enable developers to define specific workflow steps, such as code checks, artifact builds, test executions, and production deployments. Additionally, GitHub Actions can respond to webhook events, allowing workflows to be triggered by any GitHub event or third-party tool. This feature facilitates interaction with cloud services, issue updates, and notifications, thereby expanding the scope of automation.

2. Scalability GitHub Actions supports scalability by offering a wide range of pre-built actions from the Action Marketplace, which includes standards, tools, and tasks that can be smoothly integrated into workflows. Developers can leverage these pre-built actions or create custom ones to meet specific needs. The Marketplace also features numerous community-contributed actions that support a variety of tasks. Furthermore, the GitHub Actions Extension, an official VS Code extension, helps manage and execute tasks. The extension enhances development by providing syntax validation and code completion, helping developers avoid errors and ensuring a smooth workflow.

3. Integration with Azure GitHub Actions can be effectively integrated with Azure to build a powerful CI/CD pipeline for hosting microservices. This pipeline employs selective building, testing, and dynamic parallelization to accelerate value delivery. It also facilitates the creation of Docker images, which are stored in the Azure Container Registry for immediate use in testing and deployment to hosts or Kubernetes. A key feature of this pipeline is the automatic versioning of Docker images, allowing subsequent pipelines to use the same tested and built images, saving time and ensuring consistency between tested and deployed versions. Additionally, GitHub Actions enhances security by leveraging GitHub's built-in security features, such as secret management to protect workflow credentials.

Benefits of CI/CD

Early Bug Detection and Fixing: CI/CD pipelines enable developers to implement comprehensive quality checks during development and testing, creating a fast feedback loop that ensures prompt detection and resolution of emerging bugs.

Scalability: CI/CD pipelines empower Microsoft technologies, including GitHub Actions and Azure DevOps, to scale and meet the needs of both small development teams and large enterprises, in alignment with strategic goals in a dynamic business environment.

Reduced Assumptions: CI/CD pipelines replace testing assumptions with extensive, factual data, reducing cross-platform errors and enabling developers to make informed decisions during the development stage.

Improved Quality Assurance: Automation features and quality checks within CI/CD pipelines allow for smooth software deployment, overcoming challenges associated with traditional development and testing processes that often face unexpected problems and delays.

Software Health Measurability: CI/CD pipelines facilitate ongoing testing and automated deployment processes, enabling continuous monitoring of software health attributes, even for complex systems.

Lowered Overhead Costs: By detecting and fixing bugs early in the development stage, CI/CD pipelines significantly reduce overhead costs associated with repairing defects that might otherwise appear in later environments.

Enhanced Security: CI/CD pipelines enhance software security by incorporating industry-standard compliance, secrets management, and role-based access control, ensuring that security is a fundamental part of the software development process.

GitHub Actions and CI/CD pipelines offer a transformative approach to software development, streamlining workflows, enhancing scalability, and improving overall quality and security. By automating routine tasks, integrating with powerful tools such as Azure, and enabling early bug detection, these technologies not only accelerate the development process but also ensure that the final product meets the highest standards. As organizations continue to evolve in a tech-savvy era, the adoption of CI/CD practices, combined with the flexibility and power of GitHub Actions, will be essential in maintaining competitive advantage and delivering robust, reliable software solutions.

Chapter 13: Fabric's Security, Governance, and Admin

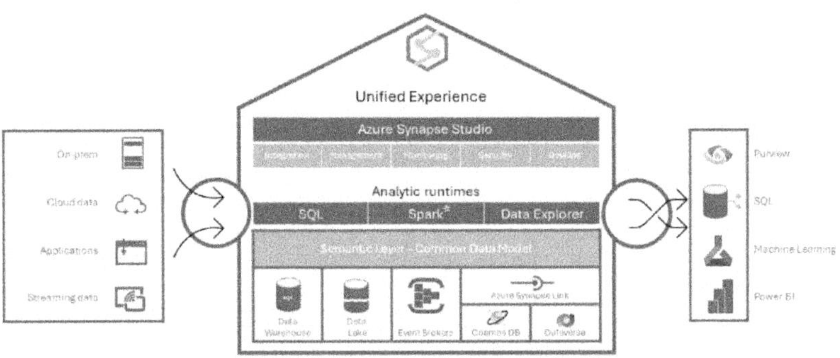

Figure 71 Unified Governance

Security in Microsoft Fabric

Ensuring the security of your data warehouse is crucial to protecting sensitive information from unauthorized access. Microsoft Fabric offers a comprehensive set of security features designed to safeguard your data warehouse, including:

- **Role-Based Access Control (RBAC):** This feature allows you to assign roles to users based on their responsibilities, ensuring only authorized personnel can access specific data.

- **SSL Encryption:** Secure communication between the data warehouse and client applications is ensured through SSL encryption, which protects data in transit from unauthorized interception.

- **Azure Storage Service Encryption:** Data is encrypted both during transfer and at rest, providing comprehensive security for your data.

- **Azure Monitor and Azure Log Analytics:** These tools enable you to monitor and track data warehouse activities, providing insights into data access patterns and potential security risks.

- **Multi-Factor Authentication (MFA):** MFA adds an extra layer of security by requiring a second form of verification, significantly reducing the risk of unauthorized access.

- **Microsoft Entra ID Integration:** This feature manages user identities and authorizations, offering a secure framework for handling access to the data warehouse.

Governance in Microsoft Fabric

Microsoft Fabric's OneLake provides centralized governance and collaboration, making data management both secure and accessible to authorized users across your organization. Governance in Fabric is managed through the admin center, which integrates seamlessly with Microsoft Purview Information Protection. This integration enables the application of sensitivity labels, helping identify and secure sensitive data throughout its lifecycle.

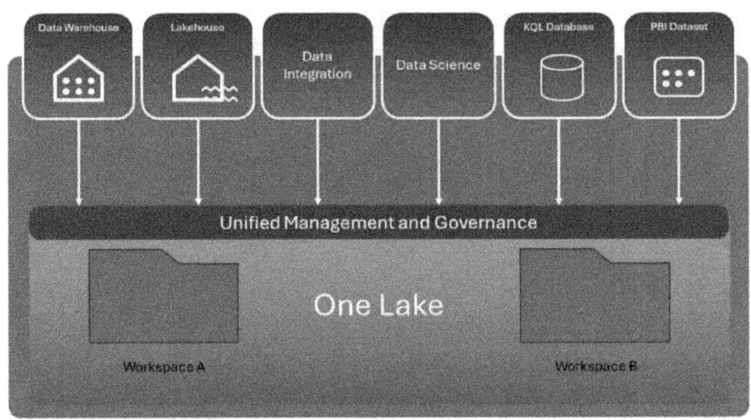

Figure 72 Governance Overview

Administrators can control groups and permissions, set up data sources and gateways, and monitor usage and performance through the admin center. Additionally, Fabric's admin APIs and SDKs enable automation of common tasks and integration with other systems, streamlining operations and ensuring efficiency.

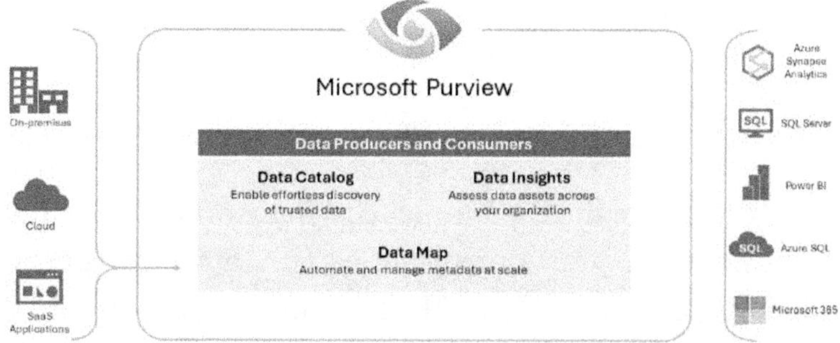

Figure 73 Microsoft Purview

Governing Data in Fabric

Administrators have access to several key governance features within Microsoft Fabric that ensure data oversight and security:

- **Content Endorsement:** This feature allows admins to mark certain Fabric items as reliable and authorized for organizational use. Endorsed content is identified with badges, helping users trust the data they rely on and ensuring the integrity of information across the organization.

- **Sensitive Data Scanning:** The scanner API enables administrators to search for sensitive data within Fabric items, including both structured and unstructured data, ensuring comprehensive data protection.

- **Data Lineage Tracking:** Data lineage allows administrators to follow the movement of data through Fabric, providing insights into the data's source, transformation, and destination. This capability enhances understanding of how data is used within the organization and supports informed decision-making.

Activating and Administering Microsoft Fabric

To fully leverage the capabilities of Microsoft Fabric, it must be activated for your organization. This typically involves collaboration with your IT department to ensure the necessary permissions are in place. The key roles involved in managing Fabric include:

- **Microsoft 365 Admin:** This role provides unlimited access to all management features, including the ability to assign roles to other users.

- **Billing Administrator:** Responsible for managing subscriptions and licenses.

- **License Administrator:** Handles the assignment and removal of licenses for users.

- **User Administrator:** Manages users and groups, as well as password resets.

Admin Tasks and Tools

In addition to the Microsoft Fabric admin portal, administrators should be familiar with various other tools to effectively manage the platform:

- **Microsoft Fabric Admin Portal:** This portal is central to managing capacities, ensuring service quality, handling workspaces, publishing visuals, and troubleshooting issues.

- **Microsoft 365 Admin Portal:** Used to manage users, groups, and licenses, and to block users from accessing Microsoft Fabric if necessary.

- **Microsoft 365 Security & Microsoft Purview Compliance Portal:** This portal oversees auditing, data classification, data loss prevention policies, and lifecycle management.

- **Microsoft Entra ID in the Azure Portal:** Configures conditional access to Microsoft Fabric resources, providing an additional layer of security.

- **PowerShell Cmdlets:** PowerShell enables administrators to manage workspaces and other aspects of Microsoft Fabric through scripting, offering a powerful tool for automating tasks and customizing the platform.

- **Administrative APIs and SDKs:** These tools allow for the development of custom admin tools, further enhancing the flexibility and functionality of Microsoft Fabric.

Microsoft Fabric: A Comprehensive Enterprise Analytics Solution

Microsoft Fabric is a robust enterprise analytics platform that covers the entire spectrum of data management, from data warehousing and engineering to data integration, science, real-time analytics, and business intelligence. Built on a SaaS model, Fabric offers a unified and streamlined solution for managing all aspects of enterprise data.

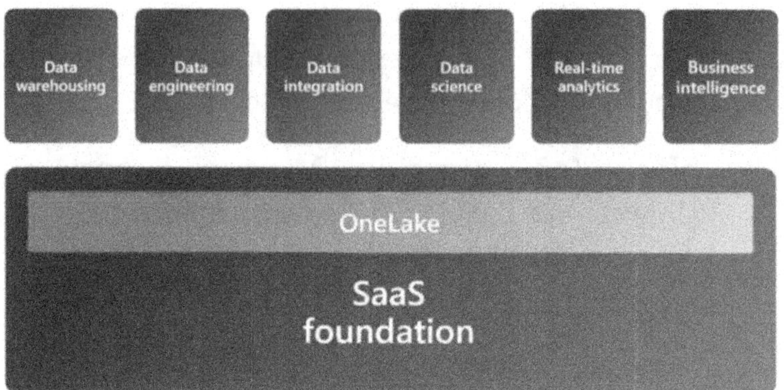

Figure 74 SaaS Foundation

By centralizing these capabilities within a single platform, Microsoft Fabric simplifies data management and provides organizations with the tools they need to harness the full potential of their data, driving informed decision-making and achieving business objectives.

Microsoft Fabric stands out as a comprehensive solution for enterprise data management, offering powerful tools for security, governance, and administration. Its robust features ensure that data is securely managed and accessible, while its governance capabilities provide administrators with the control needed to maintain compliance and trust within the organization. With its integrated approach to analytics, Microsoft Fabric enables businesses to leverage their data effectively, fostering innovation and achieving strategic goals.

Microsoft Purview and Its Integration with Microsoft Fabric

Microsoft Purview, in combination with Microsoft Fabric, forms a powerful duo within the Microsoft Intelligent Data

Platform, enabling organizations to seamlessly manage, explore, and secure their data across diverse environments. By integrating these tools, organizations can achieve comprehensive control over their entire data estate, from the source to the final Power BI report. This integration eliminates the need to rely on multiple services from different providers, offering a unified approach to data governance and security.

Unified Data Governance and Compliance with Microsoft Purview

At its core, Microsoft Purview provides robust solutions for data governance and compliance, ensuring that organizations can safeguard confidential information, assess data risks, and adhere to regulatory obligations. It supports various data environments, including Microsoft 365, on-premises, multicloud, and SaaS platforms.

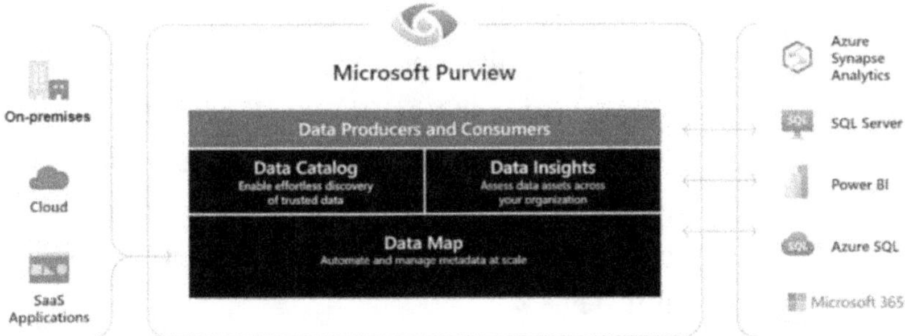

Figure 75 Unified Governance

With Microsoft Purview, organizations can:

1. **Safeguard Confidential Information**: Microsoft Purview provides advanced tools to protect sensitive data across various cloud environments,

applications, and devices. This is crucial for organizations handling large volumes of data that need to maintain privacy and security at all levels.

2. **Assess Data Risks and Ensure Compliance**: The platform offers comprehensive risk assessment capabilities, helping organizations identify potential data hazards and ensure that they comply with regulatory requirements. This is particularly important in industries with strict data protection laws, such as finance and healthcare.

3. **Gain Insight into Regulatory Compliance**: Microsoft Purview keeps organizations informed about the latest regulatory requirements, offering guidance on how to stay compliant. This proactive approach helps organizations avoid costly fines and reputational damage.

4. **Map the Entire Data Landscape**: With advanced data mapping capabilities, Microsoft Purview allows organizations to categorize their data and trace its lineage from origin to destination. This helps in understanding the flow of data within the organization, identifying potential bottlenecks or vulnerabilities.

5. **Locate Sensitive Data**: Microsoft Purview enables organizations to locate and categorize sensitive data within their environment, ensuring that it is appropriately protected and managed. This is particularly useful for organizations with complex data environments spread across multiple platforms.

6. **Provide Secure Data Access**: By creating a secure environment for data users, Microsoft Purview ensures that only authorized personnel can access critical data, reducing the risk of unauthorized access or data breaches.

7. **Understand Data Management Practices**: Microsoft Purview offers insights into how data is managed and utilized within the organization, helping to identify areas for improvement or optimization. This data-driven approach ensures that organizations can make informed decisions about their data management strategies.

Integration of Microsoft Purview with Microsoft Fabric

The integration of Microsoft Purview with Microsoft Fabric enhances the capabilities of both platforms, allowing users to manage and explore their data assets more effectively. This integration brings several key features:

1. **Microsoft Purview Data Catalog**: The Data Catalog feature allows users to view metadata related to their Microsoft Fabric items, offering a live view of the data landscape. This integration ensures that data is easily discoverable and understandable, facilitating better decision-making and data management. Users can also link their data catalog to Microsoft Fabric across multiple tenants, providing a unified view of their data assets.

2. **Microsoft Purview Information Protection**: This feature enables organizations to apply sensitivity labels to their Fabric data, ensuring that sensitive

information is protected even when exported through supported paths. Compliance administrators can track activities related to these labels through the Microsoft Purview Audit, ensuring that data security is maintained at all times. Sensitivity labels can be applied to all Fabric items, providing a consistent approach to data protection across the organization.

3. **Microsoft Purview Data Loss Prevention (DLP)**: DLP policies within Microsoft Fabric are particularly valuable for Power BI semantic models. These policies prevent the uploading of sensitive data into semantic models, identifying sensitivity labels and sensitive information types such as credit card numbers or social security numbers. DLP policies can be configured to provide policy tips to semantic model owners and alerts to security administrators, ensuring that potential risks are flagged and addressed promptly. Additionally, DLP policies can be set up to allow data owners to bypass them under certain conditions, providing flexibility while maintaining security.

4. **Microsoft Purview Audit**: User actions within Microsoft Fabric are recorded and accessible through the Microsoft Purview audit log. This feature provides detailed insights into how data is being accessed and used within the organization, helping to monitor and enforce compliance with data governance policies. Administrators can use

this information to identify potential security risks and take corrective action if necessary.

5. **Future Collaboration Between Microsoft Purview and Microsoft Fabric**: As Microsoft continues to enhance the integration between Purview and Fabric, organizations can expect even more options for monitoring and controlling their data assets. This ongoing collaboration ensures that the platforms evolve to meet the changing needs of businesses, providing a scalable and adaptable solution for data management and governance.

The Microsoft Purview Hub in Microsoft Fabric

The Microsoft Purview Hub within Microsoft Fabric serves as a centralized platform for administrators and users to manage and oversee their Fabric data assets. It provides a comprehensive view of data sensitivity and item approval, offering insights into how data is being managed within the organization. The hub also acts as a gateway to more advanced features available in the Microsoft Purview governance and compliance portals, such as Data Catalog, Information Protection, Data Loss Prevention, and Audit.

For Fabric administrators, the Purview Hub offers a detailed overview of the organization's entire data estate, allowing them to make informed decisions about data governance and management. Users, on the other hand, can access insights related to their specific data assets and explore additional features in the Microsoft Purview governance portal to enhance their understanding and management of data.

Microsoft Purview Data Map, Data Catalog, and Estate Insights

Microsoft Purview is structured around three main components, each offering unique capabilities to support data governance:

1. **Microsoft Purview Data Map**: The Data Map serves as the foundation for organizing and managing data assets within Microsoft Purview. It allows organizations to group and categorize their existing data stores, making it easier to discover and manage relevant data. The Data Map can scan data stores to collect metadata, including schemas, data types, and sensitive data classifications. This information is crucial for maintaining a comprehensive view of the data estate and ensuring that sensitive information is appropriately monitored and protected.

2. **Microsoft Purview Data Catalog**: The Data Catalog provides users with a detailed view of the metadata in the Data Map, enabling them to explore and understand their data assets more effectively. Users can see the origin of data sources, identify subject matter experts, and access data through various Azure products, such as the Azure Synapse Analytics workspace. The Data Catalog is an essential tool for users who need to find reliable and trustworthy data for their analyses and decision-making processes.

3. **Microsoft Purview Data Estate Insights**: Data
    Estate Insights offers a comprehensive overview of
    the organization's data landscape, providing key
    metrics and reports on data stewardship, catalog
    usage, asset distribution, scan performance, glossary
    classification, and sensitivity findings. These
    insights are presented in the form of charts or key
    performance indicators, helping administrators and
    security teams monitor the effectiveness of their
    data governance strategies. By leveraging these
    insights, organizations can identify areas for
    improvement and ensure that their data management
    practices align with industry standards and best
    practices.

Microsoft Purview Risk and Compliance Solutions

Risk and compliance are critical aspects of data governance, particularly in today's complex and highly regulated business environment. Microsoft Purview offers a suite of risk and compliance solutions designed to help organizations manage data risks, protect sensitive information, and adhere to regulatory requirements. These solutions include:

1. **Communication Compliance**: Ensures that all
    forms of communication within the organization
    adhere to internal policies and external regulations.
    This is particularly important in industries where
    communication must be closely monitored to
    prevent the leakage of sensitive information or the
    violation of compliance standards.

2. **Data Lifecycle Management**: Manages the entire lifecycle of data, from creation to deletion, ensuring that data is appropriately retained, archived, or disposed of according to regulatory requirements. This helps organizations avoid potential legal issues related to data retention and destruction.

3. **Records Management**: Provides a structured approach to managing records, ensuring that they are accurately maintained and easily retrievable when needed. This is essential for organizations that must comply with legal or regulatory requirements related to recordkeeping.

4. **Audit (Premium and Standard)**: Offers detailed auditing capabilities, allowing organizations to track user activities and ensure compliance with internal policies and external regulations. The audit logs provide a valuable resource for investigating security incidents or compliance breaches.

5. **eDiscovery (Premium and Standard)**: Facilitates the identification and retrieval of electronic information that may be relevant to legal cases or regulatory inquiries. eDiscovery tools help organizations manage legal risks by ensuring that relevant data can be quickly and accurately identified.

The Symbiotic Power of Microsoft 365 and Fabric

The integration of Microsoft 365 with Microsoft Fabric, enabled through Microsoft Graph, allows organizations to leverage the vast amount of data generated by Microsoft 365 applications, such as Teams, Outlook, SharePoint, and Viva Insights. Microsoft Graph provides a consistent model for accessing this data, enabling organizations to gain valuable insights into customer interactions, business workflows, security and compliance, and overall organizational efficiency.

With the introduction of Microsoft 365 Data Integration for Microsoft Fabric, organizations can now bring their Microsoft 365 data into Microsoft Fabric, creating a unified platform for data analysis. This integration allows users to work with Microsoft 365 and other data sources within a single environment, using a set of analytical tools that work together seamlessly. Microsoft Fabric's offerings, including Data Factory, Data Engineering, Data Warehousing, Real-Time Analytics, Data Science, and Power BI, all contribute to turning data into a competitive advantage.

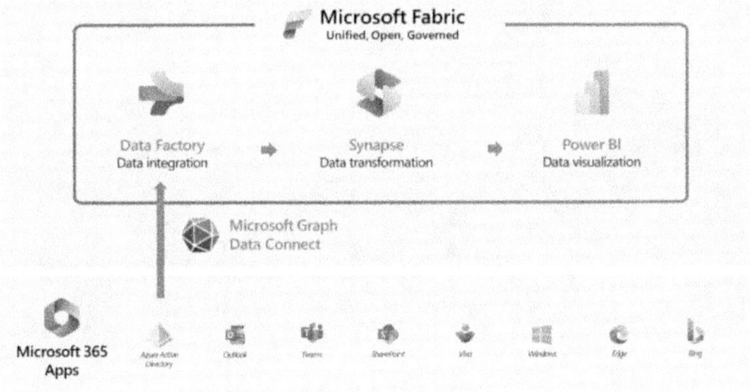

Figure 76 Fabric and O365

Microsoft Graph Data Connect provides the means to safely and efficiently transfer important Microsoft 365 datasets into

Microsoft Fabric. These datasets are prepared in three formats-basic, cleaned, and curated-offering flexibility and customization based on specific use cases or analytics scenarios. Whether organizations need raw data, normalized datasets, or custom-tailored datasets for specific applications, this integration supports a wide range of analytical needs.

How Microsoft Purview and Fabric Work Together for Enhanced Data Management

The integration of Microsoft Purview with Microsoft Fabric is designed to provide organizations with a comprehensive approach to data governance, security, and compliance, all while simplifying data management across diverse environments.

1. **Seamless Data Discovery and Classification**: The integration allows for the automatic classification and discovery of data within Microsoft Fabric, leveraging Microsoft Purview's advanced scanning and classification capabilities. This ensures that data within Fabric is not only organized and accessible but also classified according to sensitivity and compliance requirements. It streamlines the process of identifying and managing critical data assets.

2. **Consistent Policy Enforcement**: Microsoft Purview's integration with Fabric ensures that governance and compliance policies are consistently enforced across the data lifecycle. This includes applying sensitivity labels, enforcing Data Loss Prevention (DLP) policies, and ensuring that data is used in accordance with organizational standards. The result is a unified and coherent governance framework that spans the entire data estate.

3. **Enhanced Security and Compliance Monitoring**: With Microsoft Purview's comprehensive auditing and compliance tools, organizations can monitor

and analyze data usage within Fabric. This includes tracking user activities, access patterns, and potential security threats, all within a centralized platform. The ability to drill down into specific events and actions provides granular control over data security and compliance.

4. **Unified Data Catalog and Metadata Management**: Microsoft Purview's Data Catalog integration with Fabric enables users to access a unified view of their data assets, including metadata and lineage information. This not only improves data discoverability but also helps organizations maintain a consistent and accurate understanding of their data landscape. Users can easily find and utilize the data they need, with confidence in its quality and governance.

5. **Streamlined Data Operations**: By integrating Purview's governance capabilities directly into the Fabric environment, organizations can streamline their data operations. This includes automating data classification, simplifying data access controls, and providing self-service data governance tools to users. The result is a more agile and responsive data environment, where governance and security are built into everyday operations.

6. **Cross-Platform Data Management**: Microsoft Purview's ability to manage data across multiple environments-including on-premises, multicloud, and SaaS platforms-complements Fabric's data unification capabilities. This allows organizations to

manage their entire data estate from a single platform, regardless of where their data resides. The integration of Purview with Fabric ensures that data governance and security policies are applied consistently across all environments.

7. **Future-Proofing Data Governance**: As organizations grow and their data environments become more complex, the integration of Microsoft Purview and Fabric provides a scalable solution that can adapt to changing needs. Whether it's integrating new data sources, expanding data governance policies, or adopting new compliance requirements, the combined capabilities of Purview and Fabric offer a future-proof approach to data management.

Key Benefits of the Microsoft Purview and Fabric Integration

The integration of Microsoft Purview with Microsoft Fabric offers numerous benefits to organizations looking to enhance their data governance, security, and compliance capabilities:

- **Improved Data Trust**: With robust data governance and consistent policy enforcement, organizations can trust the accuracy, security, and compliance of their data. This trust is critical for making informed business decisions and maintaining regulatory compliance.

- **Increased Efficiency**: By automating many aspects of data governance and security, the integration reduces the administrative burden on IT teams, allowing them to focus on more strategic initiatives. It also empowers users with self-service tools, reducing the need for manual intervention in data management.

- **Enhanced Collaboration**: The integration fosters better collaboration between data governance teams, IT, and business users. With a unified view of the data landscape and consistent governance policies, teams can work together more effectively to achieve organizational goals.

- **Scalable and Flexible**: The integration is designed to scale with the organization's needs, supporting everything from small data environments to large, complex data estates. This flexibility ensures that the solution can grow and evolve alongside the organization.

- **Comprehensive Compliance**: With Microsoft Purview's advanced compliance tools, organizations can ensure that they meet all regulatory requirements, reducing the risk of fines and legal issues. The integration with Fabric extends these compliance capabilities across the entire data lifecycle.

- **Unified Data Governance**: The integration provides a unified approach to data governance, where policies and controls are consistently applied across all data assets, regardless of their location. This ensures that the organization's data is managed according to best practices and industry standards.

The integration of Microsoft Purview with Microsoft Fabric represents a significant advancement in data governance, security, and compliance within the Microsoft ecosystem. By combining the strengths of both platforms, organizations can achieve a comprehensive and cohesive approach to managing their data assets. This integration not only simplifies data management but also enhances the organization's ability to protect sensitive information, ensure compliance, and make data-driven decisions.

As organizations continue to navigate the complexities of modern data environments, the combined capabilities of Microsoft Purview and Microsoft Fabric provide a robust and scalable solution that can meet the demands of today's business landscape. Whether it's unifying data across multiple environments, ensuring consistent governance, or empowering users with self-service tools, the integration of Purview and Fabric offers a powerful approach to data management that is both effective and future-ready.

Chapter 14: Why Governance is Integrated into Fabric

In today's rapidly evolving world, characterized by significant environmental, technological, political, economic, and social shifts, businesses across all sectors face intensified competition. To navigate this landscape and secure a competitive edge, organizations increasingly turn to technology as a critical asset. Among the various strategies that organizations deploy, IT governance stands out as a crucial framework that can significantly influence organizational performance.

IT governance refers to the structured approach an organization adopts to ensure that its IT investments align with its strategic goals and deliver the anticipated business value. Given the substantial financial commitment many companies make toward IT, effective governance becomes essential. Despite substantial investments, results can vary widely; some organizations reap considerable benefits, while others may incur losses due to ineffective IT governance. Thus, the effectiveness of IT governance often determines whether an organization thrives or falters in its market.

Implementing robust IT governance practices is integral to aligning daily business operations with broader strategic objectives. By enhancing competitiveness and bolstering reputations through improved service quality, IT governance helps organizations expand their market share. However, the positive impact of IT governance hinges on the effectiveness of its execution. Therefore, organizations must focus on practical strategies to enhance their governance frameworks. This focus involves adopting mechanisms that steer, coordinate, and rationalize IT-related decisions.

The effectiveness of IT governance is influenced by a range of internal and external factors, which organizations must carefully consider to optimize performance. These factors play a critical role in shaping IT governance and, by extension, influence overall business productivity and progress. To achieve maximum returns from IT investments, organizations must address various domains of IT governance, including risk management, value delivery, strategic management, performance evaluation, and resource management.

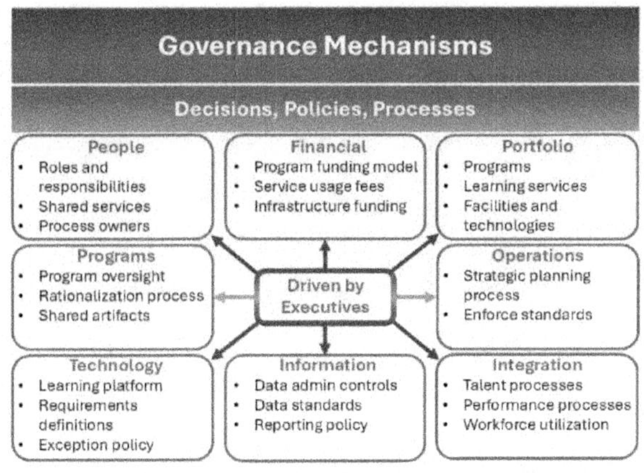

Figure 77 Governance Mechanisms

174

IT governance involves overseeing multiple accounting information systems, with the primary goal of creating a favorable environment for effective technology use. This oversight should reflect the quality of IT services, cost-effectiveness, asset utilization, and the organization's growth and flexibility. Researchers underscore the importance of measuring the value of IT, as enhancing IT governance efficiency enables senior managers and boards of directors to manage risks and achieve organizational objectives more effectively. Tools such as COBIT (Control Objectives for Information and Related Technologies) and the Balanced Scorecard are frequently cited as effective frameworks for improving IT governance.

IT Governance and Improvement Mechanisms

Influence of IT Governance on Innovation

IT governance plays a pivotal role in fostering a firm's innovation capacity. Héroux and Fortin (2016) highlight that effective resource allocation for innovation stems from well-structured IT governance mechanisms, procedures, and structures. The tools provided by IT are instrumental in driving innovation, and their effectiveness is often influenced by the governance practices in place. Good IT governance supports creativity by facilitating the acquisition, dissemination, and utilization of insights within the organization (Héroux & Fortin, 2016).

By enhancing communication across business units, IT governance creates a collaborative environment conducive to achieving shared goals. This environment can be further enriched through knowledge-sharing platforms and electronic newsletters, which help distribute relevant information efficiently. Héroux and Fortin (2016) argue that well-structured IT governance mechanisms stimulate innovation by promoting innovative thinking, inclusivity, and a collaborative corporate culture.

Moreover, IT competency is crucial in today's business environment. Héroux and Fortin (2016) assert that innovation is closely tied to the competencies of employees. Investing in IT-related training and education programs is essential for equipping staff with the skills and knowledge needed to leverage IT governance for innovation and business performance. Executive and top management must possess substantial IT skills to effectively communicate IT strategies and objectives. By prioritizing IT competency, companies can create a significant competitive advantage in the marketplace.

IT Governance Maturity

IT governance maturity is a critical aspect of strategic management that focuses on enhancing organizational capabilities through a resource-based perspective. According to Gashgari, Walters, and Wills (2017), IT governance maturity reflects an organization's ability to execute coordinated operations using its available resources to achieve desired outcomes. This model is instrumental in assessing a company's current state and identifying the best strategies for performance improvement.

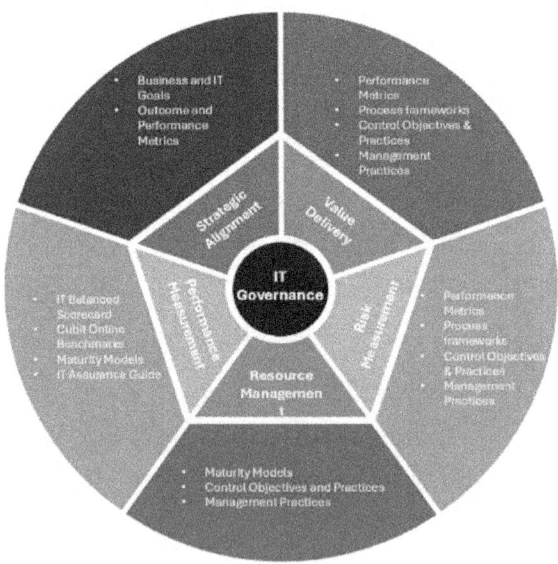

Figure 78 Governance Maturity

Gashgari et al. (2017) argue that IT governance maturity has a significant impact on business performance, productivity, and overall progress. However, they also note that some research has failed to establish a clear relationship between IT governance and business performance. Conversely, other scholars suggest that there is a considerable time lag between perceived benefits and actual IT governance maturity levels (Gashgari et al., 2017). Despite these variations, it is clear that IT governance requires a comprehensive approach covering all relevant aspects to be effective.

One key issue identified by Gashgari et al. (2017) is that many enterprises tend to overlook situational and soft governance factors, focusing instead on rigid governance structures. To enhance the scope and efficiency of IT governance, it is essential to address both hard and soft elements. Thus, understanding and improving IT governance maturity is a crucial aspect of achieving organizational success in today's dynamic business environment.

IT Governance Domains

Strategic Alignment

Strategic alignment is a fundamental domain of IT governance that ensures IT investments are in harmony with organizational goals and strategies. Bhattacharjya and Chang (2006) emphasize that effective alignment allows IT to deliver targeted business value and optimize the use of available assets and resources to achieve common objectives. The Strategic Alignment Model and the strategic alignment maturity assessment tool are key frameworks used to align internal and external procedures with IT strategy (Bhattacharjya & Chang, 2006).

Effective IT governance must ensure that IT decisions and practices are integrated with the organization's overall processes. Certain corporate operations can enhance this alignment, while others may hinder it, impacting IT department performance (Novotny, Bernroider & Koch, 2012). Factors influencing this alignment include governance styles, inclusivity, corporate culture, managerial approaches, communication, and decision-making (Wiedenhoft, Luciano, & Magnagnagno, 2017). As business performance is a collective responsibility, achieving strategic alignment is crucial for improving IT governance.

Figure 79 Strategic Alignment with Corporate Vision

Value Delivery

For IT investments to be worthwhile, they must add significant value to the organization. Lunardi, Maçada, Becker, and Van Grembergen (2016) argue that value delivery involves not only financial returns but also improvements in managerial and operational processes. Measuring the value added by IT investments can be challenging, as benefits often become integrated into daily operations. Nonetheless, effective IT governance ensures that relevant decisions are made regarding IT implementation, enhancing business value and IT productivity (Wiedenhoft et al., 2017).

Strategically evaluating IT enables organizations to uncover more opportunities, reduce risks, improve operational impacts, and achieve business objectives. Therefore, IT governance mechanisms must be employed to maximize investment returns and ensure that IT contributes positively to the organization's goals.

IT Risk Management

As organizations increasingly rely on IT, they face growing risks from external factors that can impede the achievement of corporate goals. Often, managerial focus is primarily on internal factors, neglecting the potential impact of external events. These external risks can severely damage an organization's reputation, lead to legal liabilities, and result in business losses.

The IT risk management domain addresses the protection of IT resources, operational continuity, and disaster recovery. Lunardi et al. (2016) highlight that effective risk management maintains the added value of IT investments by addressing potential risks. Top executives must understand the root causes of these risks and ensure they are controlled. Effective IT governance involves managing and mitigating these risks to safeguard strategic organizational objectives (Wiedenhoft et al., 2017). Consequently, risk management is a vital domain for improving IT governance.

Resource Management

Resource management is a crucial aspect of IT governance, focusing on ensuring that IT investments align with corporate needs and objectives. According to Lunardi et al. (2016), this domain encompasses all necessary human skills, software, hardware, and managerial processes required to translate investments into IT productivity. Resource management aims to optimize IT infrastructure and investment, ensuring proper allocation and functionality.

Effective IT governance facilitates resource management by determining the most relevant IT capabilities and supporting the formulation, implementation, and adherence to budgets, processes, and strategic plans. It also enables effective communication and decision-making regarding the allocation of funds and resources (Wiedenhoft et al., 2017). Thus, managing IT resources efficiently is an indispensable aspect of improving IT governance.

IT Performance Management

IT performance management is crucial for evaluating the value and operational impact of IT investments. It involves assessing whether IT systems meet their intended objectives and contribute significantly to business goals and needs (Wiedenhoft et al., 2017). This process examines if IT operations adhere to planned schedules, budgets, and infrastructure requirements while fulfilling user needs. It serves as a transparent tool for evaluating IT capabilities, potential risks, and overall achievements. By disclosing the performance and direction of IT investments, performance management helps assess the effectiveness of IT governance (Wiedenhoft et al., 2017). Organizations can use various mechanisms to track the progress of resources, projects, services, and strategies. The insights gained from these evaluations enable senior management to make informed decisions and implement necessary adjustments to enhance IT productivity.

Strategic Ways to Improve IT Governance

Control Objectives for Information Technologies (COBIT)

Frameworks for improving IT governance offer structured approaches to enhancing organizational IT practices. One notable framework is Control Objectives for Information and Related Technologies (COBIT). According to Rezaei (2013), COBIT provides best practices for enhancing IT governance by emphasizing the importance of implementing, controlling, measuring, and improving governance effectiveness. It is widely used globally for its auditing capabilities and standardization of high-level IT governance processes.

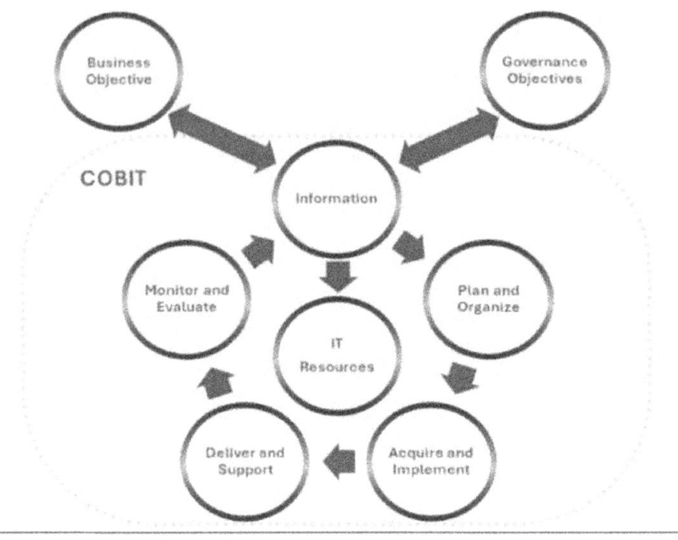

Figure 80 CORBIT Overview

COBIT is divided into four critical domains:

1. **Plan and Organize:** This initial domain focuses on developing strategic plans and tactics to integrate IT investments with business objectives. It ensures that IT supports the achievement of organizational goals (Rezaei, 2013).

2. **Acquire and Implement:** Following planning, this domain addresses the identification and resolution of functional requirements, responding to technological changes. It aims to provide effective solutions for business needs (Rezaei, 2013).

3. **Deliver and Support:** This segment emphasizes delivering support services to the IT department. It encompasses security, configuration, third-party management, and performance management to

ensure operational fluency and effectiveness (Rezaei, 2013).

4. **Monitor and Evaluate:** The final domain involves regulating governance and compliance, evaluating internal controls, and managing performance. It provides a framework for ongoing assessment and improvement of IT governance (Rezaei, 2013).

Each domain offers distinct roles, activities, and objectives that contribute to overall IT performance and governance (Zhang & Le, 2013). Organizations are encouraged to enhance their IT governance practices based on these domains to achieve their IT-related goals.

Despite the benefits of COBIT, many organizations have not fully utilized this framework. Rezaei (2013) reports that over 80% of firms do not implement COBIT in their IT operations. While the framework has shown notable improvements in IT and corporate performance for those who use it, its complexity and cost can be barriers. COBIT's extensive requirements and the need for a skilled workforce present challenges, especially for medium-sized organizations. Nevertheless, COBIT's focus on systems security and comprehensive governance remains valuable. Leaders and executives should strive to leverage COBIT's advantages to optimize their IT governance practices.

Control Objectives for Information and Related Technologies (COBIT) is widely recognized as a robust framework designed to improve IT governance and align it with business objectives. The framework aims to enhance IT effectiveness, minimize the time required to execute IT mandates, and support the achievement of corporate goals and visions. However, COBIT's complexity requires careful consideration and understanding to ensure its successful implementation and integration within an organization.

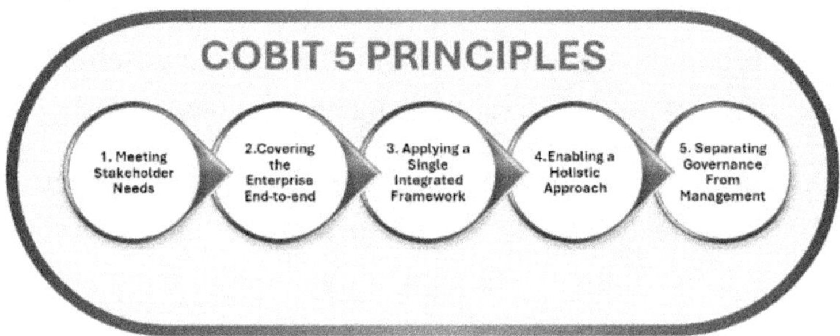

Figure 81 CORBIT 5 Principles

Objectives of COBIT

The primary objective of COBIT is to optimize the value of IT investments while managing associated risks. The framework provides a comprehensive set of guidelines and best practices to help organizations implement and manage IT governance processes. COBIT's multifaceted approach ensures that IT contributes effectively to achieving business objectives by providing a structured method for planning, executing, and evaluating IT operations.

1. **Enhancing IT Effectiveness:** COBIT aims to improve the overall effectiveness of IT systems by establishing clear objectives and performance metrics. This includes streamlining IT processes to reduce inefficiencies and ensure that IT resources are used effectively to support business goals.

2. **Reducing Execution Time:** By providing structured processes and clear guidelines, COBIT helps organizations reduce the time required to implement IT initiatives. This leads to quicker realization of benefits and a more agile IT environment.

3. **Supporting Corporate Goals:** COBIT aligns IT operations with broader business objectives, ensuring that IT investments support the organization's strategic vision and contribute to its long-term success.

4. **Complexity Management:** Due to its comprehensive nature, COBIT requires skilled and experienced personnel to avoid communication failures between IT and business units. Effective implementation of COBIT involves addressing complex processes and ensuring that all stakeholders are aligned with IT governance practices.

IT Governance Rationale Mechanisms

To enhance IT governance, organizations must understand and implement rationale mechanisms that facilitate effective communication and decision-making. According to Bianchi, Sousa, Pereira, and Hillegersberg (2017), these mechanisms involve learning and sharing strategic decisions between IT and other business units.

1. **Center of Responsibility Approach:** Many organizations use a center of responsibility model, which divides complex tasks into specialized

departments. Each department works independently but has collective responsibilities toward achieving corporate objectives. Ensuring seamless communication between executives, managers, and employees is crucial for effective IT governance.

2. **Communication Quality and Frequency:** The quality and frequency of communication between different IT stakeholders play a critical role in governance. Tactical communication, such as interactions between departmental managers and the IT department, is essential for implementing decisions effectively.

3. **Strategic Dialogue:** For optimal IT governance, strategic-level dialogue is recommended. Regular meetings between IT committees and relevant departments ensure that IT decisions align with overall business strategies. This approach enhances coordination and ensures that IT practices support the organization's goals.

4. **Board Oversight:** The governance board should oversee IT-related practices, including management, planning, performance, and evaluation. Effective oversight helps in developing performance metrics, ensuring compliance with regulations, and clarifying decision-making processes.

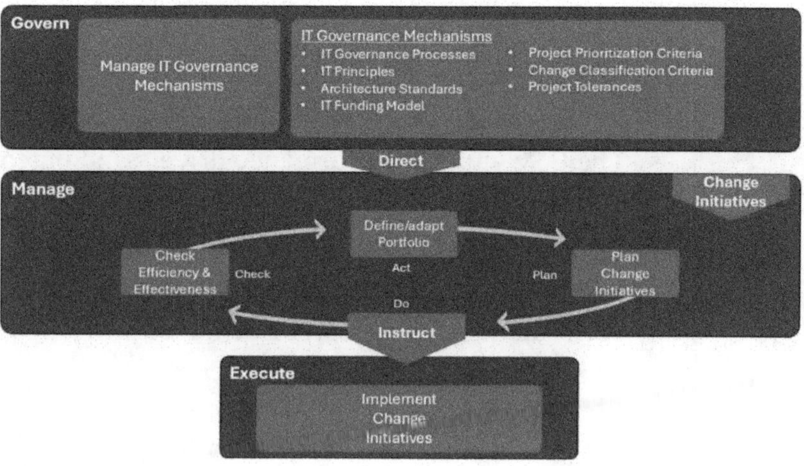

Figure 82 IT Governance Mechanism

By implementing these rationale mechanisms, organizations can improve their technological effectiveness and ensure that IT governance practices are well-aligned with business objectives.

Improving IT Governance Using the Balanced Scorecard Approach

The Balanced Scorecard (BSC) is a strategic management tool that measures organizational performance from multiple perspectives. According to Van Grembergen and De Haes (2005), the Balanced Scorecard evaluates progress beyond financial metrics, considering additional dimensions that contribute to achieving corporate objectives.

1. **Basic Principles:** The BSC approach emphasizes evaluating organizational progress through a combination of objectives, mission, and future initiatives. It focuses on cause-and-effect

relationships between performance drivers and outcomes.

2. **Performance Drivers and Outcome Measures:** The BSC utilizes performance drivers and outcome measures to assess how different aspects of the organization contribute to its overall success. This approach helps in communicating how various performance metrics can be achieved simultaneously.

3. **IT Process Relevance:** The Balanced Scorecard is particularly relevant to IT processes and functions. It evaluates IT performance from an executive management perspective, assessing its contribution to the organization's goals.

4. **Internal Users Dimension:** This dimension measures customer orientation and the efficiency of IT processes. It helps in evaluating how well IT services meet the needs of internal users and contribute to operational excellence.

5. **Future Perspective:** The future perspective of the BSC assesses organizational readiness for upcoming IT challenges and uncertainties. It improves transparency and governance by preparing the organization for future IT developments.

6. **Financial Performance:** The BSC measures the contribution of IT investments to the company's financial performance. This perspective helps in evaluating the return on IT investments and their impact on the organization's financial health.

7. **Stakeholder Dimension:** This dimension assesses the firm's status in terms of legal and ethical compliance. It ensures that IT practices adhere to relevant regulations and ethical standards.

8. **Growth and Learning:** The growth and learning segment measures areas requiring further training, learning, and development. It helps in identifying opportunities for skill enhancement and organizational growth.

Overall, the **Balanced Scorecard** approach offers valuable insights into improving IT governance. By incorporating multiple perspectives into performance evaluation, organizations can enhance their IT processes and align them more effectively with business objectives.

COBIT and the Balanced Scorecard are both effective frameworks for improving IT governance. COBIT provides a structured approach to managing IT processes, enhancing effectiveness, and supporting business goals. The Balanced Scorecard offers a comprehensive evaluation of IT performance from multiple perspectives, ensuring that IT investments contribute to organizational success. By understanding and implementing these frameworks, organizations can enhance their IT governance practices, align IT with business objectives, and achieve long-term success.

Recommendations for Enhancing IT Governance: COBIT and Balanced Scorecard Approaches

To effectively improve IT governance within an organization, both the COBIT framework and the Balanced

Scorecard approach offer valuable methodologies. Each framework provides distinct advantages that can be leveraged to align IT operations with business goals and enhance overall governance. However, the choice and implementation of these frameworks should be carefully considered based on the organization's specific needs and context.

The COBIT Framework

COBIT (Control Objectives for Information and Related Technologies) is widely recognized for its comprehensive approach to IT governance. As Batyashe and Iyamu (2016) note, COBIT is highly regarded due to its practical applicability and effectiveness in a range of organizational settings. However, it is essential to acknowledge that COBIT's complexity means it may not fit every situation without customization.

1. **Understanding and Customization:** The success of implementing COBIT relies heavily on the executive and top management's understanding of the framework's dynamics. Organizations must tailor COBIT to fit their unique operational needs and industry requirements (Rezaei, 2013). This customization process involves adapting COBIT's various processes and activities to align with the organization's IT objectives and operational context.

2. **Resource Allocation and Skilled Personnel:** To maximize the benefits of COBIT, organizations need to invest substantial resources and employ

skilled personnel. The framework includes numerous processes and mandates that, while beneficial for enhancing IT productivity, require knowledgeable professionals to manage effectively. The connection between COBIT's processes and tangible IT productivity must be clearly defined, necessitating experienced individuals who can bridge any gaps and ensure effective implementation (Rezaei, 2013).

3. **Enhancing IT Governance Domains:** COBIT is designed to improve IT governance across five key domains. By implementing COBIT, organizations can ensure that IT governance practices are integrated with overall business processes, improving alignment and performance. The framework's structured approach supports value delivery, risk management, and resource optimization, making it a robust tool for addressing various IT governance challenges (Rezaei, 2013).

   o **Value Delivery:** COBIT enhances value delivery by applying rationale mechanisms, processes, and flexible structures. It helps organizations focus on delivering tangible benefits from IT investments while managing risks associated with IT resources and operations.

   o **Risk Management:** The framework strengthens the risk management domain by addressing IT resource safety, operational continuity, and disaster recovery. COBIT provides guidelines for managing potential risks and ensuring that IT investments are protected and optimized.

o **Performance Measurement:** COBIT includes transparent instruments for assessing IT capabilities, potential risks, and achievements. This enables organizations to measure IT performance effectively and make informed decisions to enhance overall governance.

The Balanced Scorecard Approach

The **Balanced Scorecard (BSC)** is a strategic management tool that complements COBIT by providing a multi-dimensional view of performance.

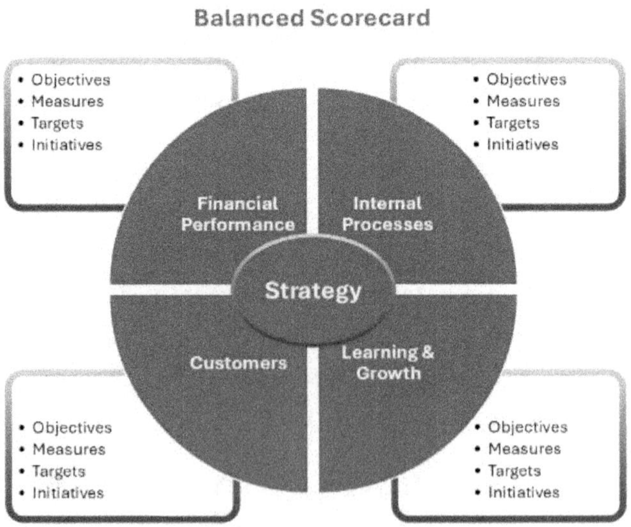

Figure 83 Balanced Scorecard

According to Van Grembergen and De Haes (2005), the Balanced Scorecard evaluates organizational progress beyond financial metrics, incorporating additional dimensions such as internal processes, customer orientation, and future readiness.

1. **Integration with COBIT:** Combining COBIT with the Balanced Scorecard approach offers a comprehensive evaluation of IT governance. The Balanced Scorecard provides insights into how IT contributes to business objectives from various perspectives, including financial performance, customer satisfaction, internal processes, and learning and growth. This integration ensures a holistic view of IT's impact on the organization and supports more informed decision-making (Borousan et al., 2011).

2. **Four Key Domains of COBIT:**

   o **Plan and Organize:** This domain focuses on controlling IT processes by aligning IT resources with business value and organizational objectives. It helps organizations manage risk, identify necessary resources, and measure performance effectively (Rezaei, 2013).

   o **Acquire and Implement:** This domain guides the acquisition and application of IT resources and measures. It ensures that IT investments add value to the organization by evaluating the effectiveness of the IT lifecycle and addressing potential future issues to protect value (Rezaei, 2013).

- o **Deliver and Support:** The Deliver and Support domain ensures that IT services and support are effectively managed to meet organizational needs. It includes aspects such as service delivery, support, and performance management.

- o **Monitor and Evaluate:** This domain involves monitoring IT performance and evaluating governance effectiveness. It provides mechanisms for assessing compliance, performance, and the overall success of IT governance practices (Rezaei, 2013).

Practical Recommendations for Implementation

1. **Tailor the Frameworks:** Organizations should customize COBIT and the Balanced Scorecard to fit their specific industry needs and operational contexts. This involves adapting the frameworks' processes, metrics, and guidelines to align with the organization's goals and challenges.

2. **Invest in Training and Resources:** Adequate investment in training and resources is crucial for the successful implementation of both frameworks. Organizations should ensure that their personnel are well-versed in COBIT and the Balanced Scorecard methodologies to maximize their benefits.

3. **Regular Review and Adaptation:** IT governance practices should be regularly reviewed and adapted based on evolving business needs and technological advancements. Continuous assessment helps organizations stay aligned with their objectives and address any emerging challenges effectively.

4. **Foster Communication and Collaboration:** Effective communication and collaboration between IT departments and other business units are essential for successful implementation. Regular meetings and strategic dialogues can ensure that IT decisions are well-integrated with overall business strategies and objectives.

By following these recommendations, organizations can leverage the strengths of COBIT and the Balanced Scorecard to enhance their IT governance practices. Both frameworks offer valuable insights and tools for improving IT performance, aligning IT with business goals, and achieving long-term success.

Addressing complex technological challenges within organizations can be effectively managed through the adoption of the COBIT and Balanced Scorecard models. These frameworks significantly enhance IT governance, which is crucial for maintaining competitiveness and relevance in today's rapidly evolving market landscape. By leveraging these models, companies can better navigate the intricate terrain of IT management and drive organizational performance.

Importance of IT Governance

In the contemporary business environment, technology plays a pivotal role in shaping a company's market position and operational efficiency. Effective IT governance is essential for ensuring that technological investments align with organizational goals and deliver the expected business value. IT governance encompasses several critical domains: strategic alignment, value delivery, risk management, resource management, and performance management. Each of these domains contributes to an organization's success or failure by influencing how IT supports and drives business outcomes.

Strategic alignment ensures that IT initiatives are in sync with the overall business strategy. Value delivery focuses on maximizing the return on IT investments. Risk management addresses potential threats to IT systems and data. Resource management optimizes the use of IT resources, and performance management assesses the effectiveness and efficiency of IT operations. Understanding the impact of these factors on IT governance and overall business performance is vital for enhancing productivity and achieving progress.

COBIT Framework

The COBIT (Control Objectives for Information and Related Technologies) framework is a comprehensive tool designed to improve IT governance through best practices and structured processes. It includes four key domains:

1. **Plan and Organize:** This domain helps IT departments manage and control IT processes by aligning IT resources with business objectives. It

focuses on planning and organizing IT resources to support strategic goals.

2. **Acquire and Implement:** This segment guides the acquisition and implementation of necessary IT resources, mechanisms, and measures. It ensures that IT investments are effectively utilized and contribute to organizational goals.

3. **Deliver and Support:** This domain emphasizes the delivery of IT services and support, ensuring that IT operations meet organizational needs and performance standards.

4. **Monitor and Evaluate:** This segment involves monitoring IT performance and evaluating the effectiveness of IT governance practices. It ensures that IT systems and processes are functioning as intended and delivering value.

Despite its benefits, COBIT can be complex and may require customization to fit specific organizational needs. Effective implementation of COBIT necessitates a deep understanding of its components and careful adaptation to the organizational context. Despite these challenges, COBIT remains an invaluable tool for improving IT governance and enhancing corporate performance.

Balanced Scorecard Approach

The Balanced Scorecard (BSC) is another essential tool for evaluating and improving IT governance. It provides a multi-dimensional approach to performance measurement and management, focusing on three main elements:

objectives, mission, and future initiatives. The Balanced Scorecard evaluates IT performance from various perspectives:

- **Objectives:** It measures how well IT supports organizational goals and strategic objectives.

- **Mission:** It assesses IT's role in fulfilling the company's mission and vision.

- **Future Initiatives:** It evaluates IT's preparedness for future challenges and opportunities.

The Balanced Scorecard is relevant and applicable to IT processes and functions, offering insights into how IT investments contribute to business performance and strategic objectives.

Combining COBIT and Balanced Scorecard

When used together, COBIT and the Balanced Scorecard provide a comprehensive framework for improving IT governance. COBIT's structured approach to managing IT processes and risks complements the Balanced Scorecard's multi-dimensional performance measurement. The integration of these models enables organizations to:

- **Enhance IT Process Control:** COBIT's Plan and Organize domain supports the management of IT processes, while the Balanced Scorecard evaluates their effectiveness in achieving business objectives.

- **Optimize IT Resource Utilization:** COBIT's Acquire and Implement domain guides the acquisition and application of IT resources, while the Balanced Scorecard assesses the alignment of these resources with strategic goals.

- **Ensure IT Value Delivery:** The Monitor and Evaluate domain of COBIT ensures that IT operations deliver value, which is further measured by the Balanced Scorecard's performance metrics.

In summary, adopting the COBIT and Balanced Scorecard approaches offers organizations a robust framework for enhancing IT governance. COBIT's focus on managing IT processes, resources, and performance aligns well with the Balanced Scorecard's emphasis on evaluating IT's contribution to business objectives. By integrating these models, organizations can improve IT governance, drive business performance, and achieve their strategic goals. Companies should consider implementing both COBIT and the Balanced Scorecard to optimize their IT governance practices and ensure that their technological investments yield tangible benefits.

Chapter 15:
The Impact of IT Policy Framework

Information technology (IT) has profoundly transformed the modern world, permeating almost every aspect of daily life. The rapid integration of IT into various sectors has underscored the need for policies and regulations to govern its use, ensuring that technology is applied responsibly and securely. These policies, while sometimes expensive and resource-intensive to develop and implement, are essential in maintaining the integrity of information systems and protecting users from various forms of cybercrime and misuse. With the increasing reliance on computers, safeguarding both personal and organizational data has become a top priority.

The misuse of information technology is an escalating concern. Cybercrime is on the rise, with individuals and groups exploiting vulnerabilities in systems for personal gain, whether by attacking businesses, hacking personal accounts, or spreading misinformation. As technology continues to advance, so do the tactics employed by cybercriminals. This makes it imperative for IT policies to evolve rapidly to keep up with emerging threats. Cybersecurity experts are constantly developing new rules, regulations, and software solutions to enhance the security of IT systems, but the fast-paced nature of technological advancements often renders these solutions obsolete sooner than anticipated.

One of the major benefits of establishing and following robust IT policies is the protection they provide. These guidelines ensure that users are aware of their obligations and responsibilities when using IT systems, promoting a safer and more efficient digital environment. Policies also ensure that organizations make informed decisions about the types of IT systems they invest in and the security measures they adopt. In a world where significant transactions and operations are carried out through computers, it is vital to protect these systems from misuse.

Figure 84 Policies and Principles

The Importance of IT Policies

Investments in IT infrastructure, whether in the form of software, hardware, or personnel, can be substantial. Therefore, it is crucial for organizations to have policies in place that ensure the proper and effective use of these investments. IT policies vary widely depending on the organization's specific needs, but they generally cover the safe and ethical use of technology, the protection of sensitive information, and the prevention of unauthorized access. Without these policies, the risk of financial loss, data breaches, and reputational damage increases significantly.

IT policies also provide clear guidelines for employees, ensuring that everyone within an organization understands how to use technology in a way that aligns with the company's objectives. By promoting best practices, these policies help improve overall efficiency, reduce the risk of human error, and create a more secure working environment. Furthermore, they support compliance with legal requirements and industry standards, which is particularly important in sectors such as finance, healthcare, and government.

Acceptable Use Policies

An Acceptable Use Policy (AUP) outlines the appropriate and inappropriate use of an organization's IT resources. It is designed to protect both the user and the system by ensuring that all parties honor the rights and privileges of fellow users and the integrity of IT infrastructure. Violations of an AUP often result in serious consequences, including disciplinary action or termination of access rights.

At its core, an AUP ensures that IT systems are used for the common good. Activities that negatively impact other users-such as unauthorized access to personal data, spreading malicious software, or engaging in illegal activities-are strictly prohibited. Users are responsible for maintaining the security of their own credentials, such as passwords, and must refrain from sharing sensitive information. The improper use of IT systems, including violating others' privacy, sharing copyrighted material without permission, or distributing offensive content, is not tolerated.

For example, the introduction of malware, viruses, or unauthorized programs into a system is strictly forbidden, as these actions compromise the safety and functionality of the network. Furthermore, sharing offensive or inappropriate material, whether it is of a sexual nature or intended to spread harmful misinformation, is prohibited. The AUP underscores the importance of each individual taking responsibility for their own actions and ensuring they do not harm others in the digital environment.

Access Control Standards

Another critical aspect of IT security is the Access Control Standards Policy. This policy governs who has the right to access certain systems and data, and what actions they are permitted to take within those systems. There are two main types of access control: physical and logical.

Physical access control limits who can enter facilities where IT equipment is housed, ensuring that only authorized personnel have direct contact with critical hardware. Logical access control, on the other hand, restricts digital interactions with IT systems, such as login credentials, network access, and database management. Both forms of control are essential to maintaining the security of IT infrastructure.

To implement logical access control, users are typically required to identify themselves through a login process. Credentials such as passwords, access cards, and biometrics may be used to verify a user's identity. Once authenticated, users are granted varying levels of access depending on their role within the organization. For example, system administrators may have full access to modify and manage IT systems, while regular employees may be limited to using specific software for their job functions.

Strong password policies are a fundamental part of access control. Passwords must meet certain length and complexity requirements, and users should be prompted to change them regularly to reduce the risk of compromise. In addition, unauthorized access to confidential information is strictly prohibited. Sharing sensitive data without appropriate authorization or accessing systems beyond a user's designated permissions are serious violations.

In some cases, remote access to IT systems is necessary, particularly for users working from home or in different locations. This access must be carefully managed, with secure methods such as virtual private networks (VPNs) or encrypted communication channels to protect the integrity of the data being transmitted.

Periodic Review and Maintenance

To ensure that IT policies remain effective, regular reviews and updates are necessary. The fast pace of technological advancement means that policies can quickly become outdated, leaving organizations vulnerable to new threats. Periodic audits of IT systems, coupled with updates to security protocols, are essential to maintaining a safe and functional computing environment. This includes updating software, revising access control protocols, and reinforcing user education to keep everyone informed about the latest security measures.

Information Technology Planning Policy

An effective Information Technology (IT) planning policy provides a structured approach for acquiring, implementing, and managing IT systems within an organization. The primary goal is to align IT investments with the organization's strategic goals, ensuring that technology serves the overall mission and objectives. For this alignment to occur, there must be an integrated approach where capital planning and investment controls enhance organizational performance and adaptability to new technologies (Peltier, 2016).

A robust IT planning policy requires continuous assessment of current technologies and future needs. Analytics systems should be deployed to evaluate how well the IT policies meet emerging demands and ensure alignment with organizational objectives. Regular feedback from stakeholders-including management, employees, and external partners-should be incorporated to make necessary adjustments. Additionally, organizations must review existing and emerging best practices in the technology field to remain competitive.

Cost-benefit analysis is critical to IT planning. The economic implications of acquiring and managing new systems must be assessed against the projected benefits, such as increased efficiency, improved data security, and enhanced productivity (Sentft, Gallegos, & Davis, 2016). Establishing metrics like critical success factors and key performance indicators allows an organization to monitor the effectiveness of its IT investments over time.

Security planning is also a vital aspect of IT policy. Organizations must prepare detailed security plans that address the confidentiality, integrity, and availability of data. These plans should be regularly updated to account for the rapid evolution of technology and emerging cyber threats. Evaluation of security protocols should be continuous, with regular feedback loops integrated into the planning process to address weaknesses and improve system safety (Tirgari, 2012).

Hardware and Software Acquisition Standards Policy

Acquiring the right hardware and software is essential for maintaining operational efficiency and system security. Computer hardware, including desktops, laptops, printers, and peripheral devices, require periodic upgrades or replacements as components wear out or become outdated. However, before any acquisition, organizations should consult their IT department to ensure the hardware or software being purchased aligns with existing systems and supports current operational needs (Peltier, 2016).

The IT department must verify that new hardware and software are compatible with the organization's technical infrastructure. This process includes understanding the technical specifications, safety protocols, and auxiliary resources needed for the smooth integration of new systems. Any non-conforming or incompatible systems should be disregarded.

Vendors play a crucial role in the acquisition process. Only certified vendors with proven technical expertise should be engaged to provide new systems or offer support. Vendors must also demonstrate the system's functionality through a thorough presentation before purchase to ensure it meets the organization's requirements (Sentft, Gallegos, & Davis, 2016).

Data protection is another critical aspect of hardware and software acquisition. If any organizational data is required for software development or testing, only necessary information should be provided. Written agreements must outline how data will be used and stipulate conditions for termination in case of misuse. Software pricing should be based on system configurations and the specific components it will interact with (Tirgari, 2012).

Electronic Communications Policy (Email, Social Media, etc.)

Electronic communication, including email and social media, has become an indispensable tool for both personal and business interactions. An organization's electronic communication policy should be comprehensive, addressing appropriate usage, security, and the protection of sensitive information across all platforms.

Email and social media guidelines must clearly define acceptable content sharing and behavior. Sharing inappropriate, offensive, or unauthorized content through organizational email systems is strictly prohibited. Users must also be mindful of security risks, such as sharing personal or sensitive company information with unauthorized parties (Sentft, Gallegos, & Davis, 2016). Emails should always be encrypted when sending confidential information to prevent interception by malicious actors.

Social media usage policies should emphasize the need for privacy and security. Users should be aware of the dangers of oversharing personal information, which could expose them to cyberattacks, identity theft, or social engineering schemes. To safeguard vulnerable users, such as children, organizations should implement monitoring tools and filtering software to screen inappropriate content and prevent exposure to harmful information.

With the rise of cyberbullying, harassment, and the misuse of online platforms, organizations must also address the ethical and moral aspects of electronic communication. Employees should be educated on the importance of using respectful language and avoiding harmful behavior, both in professional and personal online interactions. Installing antivirus software, firewalls, and intrusion detection systems further ensures that organizational networks remain secure from external threats (Caliz, Samaniego, & Caliz, 2016).

Lastly, organizations should establish internet usage policies to clarify what types of online activity are acceptable and which are prohibited. These policies should limit access to unsafe websites and ensure that users do not engage in activities that could compromise system security. Filtering software is a key tool in enforcing these policies, helping to block malicious content and prevent employees from inadvertently downloading harmful files or accessing inappropriate material (Tirgari, 2012).

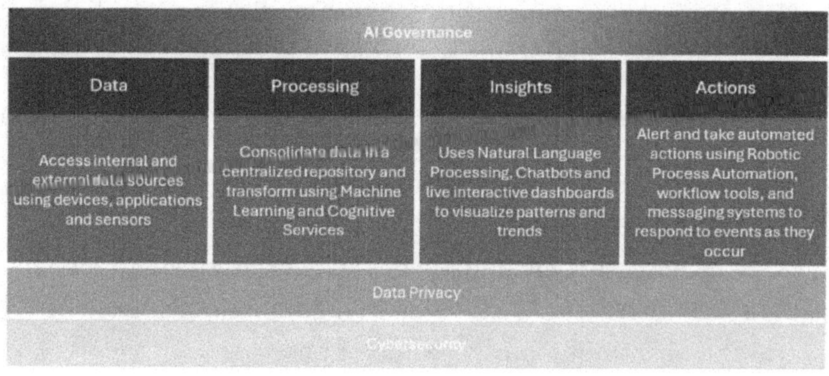

Figure 85 AI Governance Example

Data Governance Policy

A data governance policy is a formal set of guidelines that ensures the proper management, protection, and use of an organization's digital information. The primary goal of such a policy is to ensure that data is accessible, accurate, consistent, and secure, while clearly defining the roles and responsibilities for data management within the organization (Safa, Von Solms, & Furnell, 2016). While data governance policies are documented, they remain flexible to accommodate changes as technological and organizational needs evolve.

At its core, a data governance policy is essential for ensuring data quality and integrity throughout the organization. It outlines the rules and procedures for how data is collected, stored, used, and shared. This policy is critical in aligning data management practices with the organization's strategic goals, supporting efficient decision-making, compliance with regulatory requirements, and mitigating risks related to data breaches and misuse.

One of the key components of a data governance policy is **data access management**, which defines who has permission to access, modify, or share organizational data. While employees may need access to specific datasets to perform their tasks, safeguarding sensitive information is crucial. Access should be role-based, ensuring only those with the necessary clearance can view or modify confidential data (Peltier, 2016). This helps in maintaining data security without obstructing business operations.

Additionally, the policy should cover **data usage**, ensuring data is utilized only for its intended purpose and in compliance with organizational guidelines and legal regulations. Misuse or unauthorized sharing of data can lead to significant legal, financial, and reputational risks. To further mitigate risks, **data stewards** -individuals responsible for managing and overseeing data- are typically designated within the organization. These stewards ensure that data is used responsibly, updated regularly, and remains accurate.

Data integrity is another critical aspect of the governance policy, ensuring that data remains consistent, accurate, and reliable throughout its lifecycle. The policy should include mechanisms for validating data and maintaining consistency across different platforms and systems. As data is a vital asset for decision-making, maintaining its integrity is crucial for the organization's overall success.

Regular **audits and evaluations** of data management practices are essential for ensuring compliance with the governance policy. Continuous monitoring and feedback from stakeholders should inform updates to the policy, ensuring that it remains aligned with both organizational needs and external regulations.

Overall, the data governance policy provides the foundation for a well-organized, secure, and efficient data management system that supports the organization's broader business objectives (Sentft, Gallegos, & Davis, 2016).

Electronic Signature Policy

An **electronic signature policy** outlines the protocols and standards for using electronic signatures within an organization. This policy is essential for ensuring the legality, security, and consistency of electronically signed documents and agreements, transforming traditional paper-based processes into faster, more efficient workflows. By adopting an electronic signature policy, organizations can streamline operations, reduce reliance on physical documents, and expedite approval processes (Peltier, 2016).

A robust electronic signature policy should comply with relevant local and international laws, such as the U.S. **Electronic Signatures in Global and National Commerce Act (ESIGN)** and the **European Union's eIDAS regulation**, ensuring that electronic signatures have the same legal validity as handwritten signatures. Compliance with these regulations guarantees that electronic signatures are legally binding, enforceable, and secure in any legal or business transaction (Caliz, Samaniego, & Caliz, 2016).

The policy should detail the procedures for verifying the authenticity of electronic signatures. This includes encryption methods, digital certificates, and multi-factor authentication to ensure that only authorized personnel can sign documents and that the signature is tied to a verifiable identity. By leveraging these security measures, organizations can protect sensitive information and reduce the risk of fraud or unauthorized access to critical documents.

In addition to security, the electronic signature policy should specify the types of documents eligible for electronic signatures and the conditions under which electronic signatures can be used. For example, certain high-value transactions or legal contracts may require more stringent verification processes, while everyday internal approvals may have more lenient requirements.

The policy should also include guidelines for data privacy and security, ensuring that signer information is encrypted and only accessible to authorized personnel. In cases where business transactions involve third parties, the electronic signature policy must ensure that signatories' personal data is protected in compliance with privacy regulations such as **GDPR** or **HIPAA**.

Lastly, the policy should address situations where paper-based signatures may still be necessary and provide a framework for integrating these with digital workflows. Continuous evaluation of both the legal and business requirements for electronic signatures ensures that the policy remains relevant and effective as organizational needs evolve (Sentft, Gallegos, & Davis, 2016).

Change Management Policy

A **change management policy** is a structured approach to managing alterations or updates to an organization's IT systems, ensuring that changes are implemented smoothly, efficiently, and with minimal disruption to business operations. The primary goal of this policy is to mitigate risks associated with changes to critical systems and processes, such as service interruptions, data loss, or security vulnerabilities (Sentft, Gallegos, & Davis, 2016).

The change management policy outlines the process for proposing, reviewing, testing, and implementing changes to IT systems. It includes detailed guidelines for documenting each step of the process to ensure transparency and accountability. All proposed changes should be evaluated based on their potential impact on the organization's operations, security, and compliance with industry standards.

Change requests must be submitted through a formal process, typically involving the completion of a change request form that specifies the nature of the change, the systems affected, and the anticipated outcomes. The form must also include a risk assessment, detailing any potential hazards associated with the change and how these risks will be mitigated. Once submitted, the request is reviewed by a **change management board** or designated authority, which evaluates the request's feasibility and necessity (Safa, Von Solms, & Furnell, 2016).

Once approved, changes are subjected to thorough testing in a controlled environment to verify their effectiveness and compatibility with existing systems. This testing phase is critical to ensuring that any potential issues are identified and resolved before the change is implemented in the live environment.

The change management policy also includes procedures for **communication and training**, ensuring that employees and other stakeholders are informed of any upcoming changes and how these changes will impact their daily activities. This reduces resistance to change and ensures that everyone is prepared for the transition.

In cases where the change involves modifying a **firewall** or other security measures, specific protocols must be followed. For example, any changes to firewall settings must be carefully documented, and the firewall change request form must be completed to track the modifications. This ensures that security standards are maintained, and any future issues can be traced back to specific changes (Caliz, Samaniego, & Caliz, 2016).

Continuous monitoring and post-implementation review are essential components of the change management policy. After a change is implemented, its performance should be assessed to ensure that it delivers the expected benefits and that no new issues have arisen. Regular audits and evaluations ensure that the change management policy remains effective and that the organization can adapt to evolving technological and business needs.

Media Protection and Disposal Policy

The **Media Protection and Disposal Policy** is designed to safeguard organizational data by ensuring that sensitive information is protected until it is officially released through authorized channels. This policy encompasses controls for both electronic and physical media that house sensitive data, aiming to prevent unauthorized access and ensure secure handling throughout the data lifecycle (Peltier, 2016).

Media Protection: The policy mandates that all electronic media-such as memory devices in laptops and desktops, portable digital storage devices, and digital storage systems-be stored securely within controlled physical environments. Access to these media should be restricted to authorized personnel only. Computers and laptops must be logged off or shut down when not in use to prevent unauthorized access (Caliz, Samaniego, & Caliz, 2016). Similarly, printed documents containing sensitive information must be stored in secure areas accessible only to employees whose roles require access to this data.

Media Disposal: Proper disposal of media is crucial for protecting organizational knowledge and preventing data breaches. The policy dictates that all electronic media must be overwritten at least three times before disposal or release for reuse. Detailed documentation of the procedures used for data sanitization or destruction must be maintained by the organization (Sentft, Gallegos, & Davis, 2016). Physical media should be securely disposed of when no longer needed, to prevent unauthorized access and potential leaks of confidential information. Employees who fail to follow proper disposal protocols, or who dispose of media without supervisor approval, will face penalties. This ensures adherence to established procedures and maintains the integrity of the organization's data security efforts.

Security Awareness and Training Policy

The **Security Awareness and Training Policy** emphasizes the importance of comprehensive training to enhance data security across all levels of the organization. It addresses the misconception that data security is solely the responsibility of the IT department, highlighting that lapses by any employee can expose the organization to cyber threats (Safa, Von Solms, & Furnell, 2016). This policy ensures that all employees understand their role in maintaining security and are equipped with the knowledge to protect sensitive information.

Training Levels:

1. **General Awareness**: All employees receive basic training on data security principles, including recognizing phishing attempts and handling sensitive data responsibly.

2. **Intermediate Security Awareness**: Managers receive additional training focused on overseeing security practices and managing data protection at the departmental level.

3. **Advanced Security Awareness**: IT personnel undergo in-depth training on technical aspects of

data security, including advanced threat detection and incident response strategies.

Training Components: The policy requires training on various aspects of data security, including:

- Proper password practices, including strength, validity, and length requirements.

- Identification and handling of sensitive data areas within the organization.

- Procedures for securely disposing of and storing paper-based documents.

- Protocols for reporting and responding to suspected cybersecurity incidents.

The aim of the policy is to foster a culture of security awareness, enabling employees to recognize and mitigate both physical and cyber threats. By identifying and addressing the organization's vulnerabilities, the policy supports proactive rather than reactive measures to data security (Caliz, Samaniego, & Caliz, 2016).

Chapter 17: Case Study- Healthcare use of Onelake

For many years, organizations across various sectors struggled with unintegrated data systems, leading to significant inefficiencies and inconsistencies. It became clear that without a cohesive view of organizational data, making informed decisions was nearly impossible. This recognition spurred the development of data warehousing-a process designed to integrate disparate data systems within an organization. Data warehouses were established to serve as decision support systems (DSS), enabling more effective data analysis and informed decision-making (Padmanabhan & Patki, 2016). Today, the integration of data has become a fundamental necessity for organizations seeking to enhance their decision-making capabilities.

In particular, the healthcare sector, known for being one of the largest and most information-intensive industries globally, has been slower to adopt data warehousing compared to other sectors such as transportation, manufacturing, and finance. The unique nature of healthcare data, which differs significantly from data in other industries, posed challenges for effective integration.

However, as the need for integrated data in healthcare became increasingly apparent, the sector began to embrace advanced decision support systems to enhance operations and patient care.

Previous Data Warehousing Gaps in Healthcare

Despite the evident need for data warehousing, the healthcare industry has historically lagged in leveraging these systems compared to other sectors. The primary challenge lies in the distinct nature of data in healthcare, which is fundamentally different from the data types prevalent in industries like transportation or business. For example, in the transportation sector, data might include customer reservations, cargo delivery schedules, and flight timings-essentially numerical and transactional in nature. In contrast, healthcare data encompasses a wide range of patient interactions, such as outpatient visits, inpatient care, emergency responses, and various treatment procedures. This unique dataset is often not repetitive or standardized, making integration more complex (Visscher et al., 2017).

Moreover, healthcare data is predominantly textual rather than numerical. The interactions between patients and healthcare providers often involve verbal communication, which is documented in text form. For example, hospitals record patient histories and treatment plans in narrative formats, while nurses document procedures and observations through written notes (Inmon, 2005). Traditional data warehousing systems, which were designed primarily for numerical data, struggled to accommodate the rich, unstructured nature of healthcare information. As a result, the healthcare sector could not fully capitalize on the benefits of decision support systems that were tailored to numerical datasets.

However, advancements in technology and a growing understanding of the healthcare data landscape have led to the development of specialized data warehousing solutions capable of integrating textual information that's been obtained from images, x-rays, and other object data. This evolution has allowed healthcare organizations to effectively aggregate and analyze data from various sources, facilitating a more comprehensive understanding of patient care. Healthcare professionals from diverse specialties-such as pediatrics, epidemiology, gynecology, cardiology, and orthopedics-can now contribute their unique datasets to a centralized data warehouse, enriching the information available for analysis and decision-making.

The adoption of the Onelake Data Warehouse exemplifies this shift in the healthcare industry. By integrating both textual and numerical data, Onelake enables healthcare providers to leverage a unified platform for decision support. This integrated approach not only improves patient outcomes but also streamlines operational efficiencies, allowing healthcare organizations to respond more effectively to the complex needs of their patients.

As healthcare continues to embrace data warehousing solutions like Onelake, the potential for enhanced decision-making and improved patient care becomes increasingly apparent. The availability and integration of diverse data types will empower healthcare organizations to make informed decisions based on comprehensive insights, ultimately leading to a more effective and responsive healthcare system.

Initiatives Made Possible by Decision Support Systems (DSS)

Textual data warehouses have revolutionized the healthcare industry, enabling the availability, effective management, and utilization of data within hospitals and the broader medical field. This transformation has allowed healthcare organizations to benefit from data warehousing in ways similar to other industries, significantly improving healthcare delivery worldwide (Silver et al., 2011). By integrating data through Decision Support Systems (DSS), the healthcare sector has developed clearer insights across various dimensions of patient care and operational efficiency. As a result, several key initiatives have emerged, including proactive treatment programs, chronic disease monitoring, and the development of population health decision support systems.

Proactive Treatment Programs

In the business realm, automation has streamlined various processes, enhancing operational efficiency. Similarly, the healthcare sector has embraced automation through DSS to improve patient care. Automated systems now facilitate the sending of health maintenance messages to patients, reminding them of important health-related tasks. Advanced DSS have been created to distribute critical data across medical applications, including hospital billing systems, ultimately leading to enhanced healthcare quality and reduced medical costs through preventive care.

Proactive treatment programs empower patients to engage more actively in their health management, positively impacting their outcomes and overall satisfaction. For example, sending reminders for upcoming tests, visits, immunizations, or even updates on medical reports keeps patients informed and proactive about their health (Visscher et al., 2017). This increased communication leads to a higher percentage of patients participating in their care, thus enhancing the quality of healthcare services.

One particularly compelling application of proactive programs is in the area of childhood immunization. Pediatricians administer vaccinations to protect children from regional health threats while managing schedules for potentially hundreds of patients. Electronic Health Records (EHR) systems equipped with DSS automatically remind healthcare providers of upcoming immunizations and track patients' medical histories. However, if children miss their appointments, the opportunity for timely immunization is lost. DSS ensures that both doctors and patients receive notifications about upcoming visits, significantly improving immunization rates. By generating printouts of patients due for vaccinations-including their contact details, medical histories, and demographics-DSS eliminates errors and enhances the connection between patients and their healthcare providers.

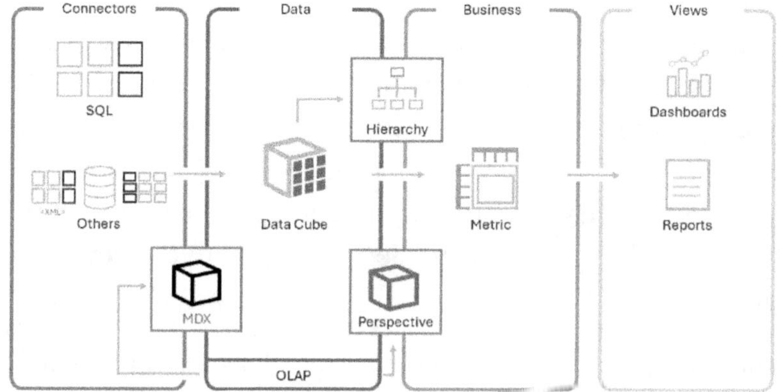

Figure 86 Online Analytical Processing (OLAP) Overview

Monitoring Chronic Diseases

Chronic diseases present complex challenges, with conditions like obesity, cancer, and diabetes being among the most pressing health concerns today. Effectively managing and reporting data related to these chronic conditions requires robust systems that ensure high-quality data reporting to improve research outcomes and patient care. Diabetes, in particular, poses significant risks if not closely monitored, as patients may develop other preventable diseases without proper oversight. DSS play a crucial role in identifying at-risk individuals and preventing the onset of chronic conditions through timely reminders for necessary tests and interventions.

Data warehouses serve as essential surveillance systems for chronic diseases, offering deeper insights into behavioral risks associated with these conditions. By leveraging personal information, DSS can identify individuals at risk, their geographic locations, and relevant health data.

Behavioral risks are analyzed based on various demographics, such as age, marital status, and past health records. This comprehensive analysis enhances the understanding of the links between specific risk behaviors and chronic diseases, such as mental health issues, cancer, and obesity. By identifying major contributing factors, healthcare providers can develop effective preventive strategies.

The data generated by DSS on chronic diseases is invaluable not only for healthcare professionals but also for government entities and public health organizations. This data informs decision-making regarding healthcare investments targeted at specific diseases, ultimately fostering a more effective allocation of resources.

The initiatives made possible by DSS and textual data warehouses are transforming the healthcare landscape. Through proactive treatment programs and improved monitoring of chronic diseases, healthcare providers can enhance patient outcomes, streamline operations, and make more informed decisions. By embracing advanced data management systems, the healthcare industry is poised to tackle the complexities of modern medical care, ultimately leading to better health outcomes for communities worldwide.

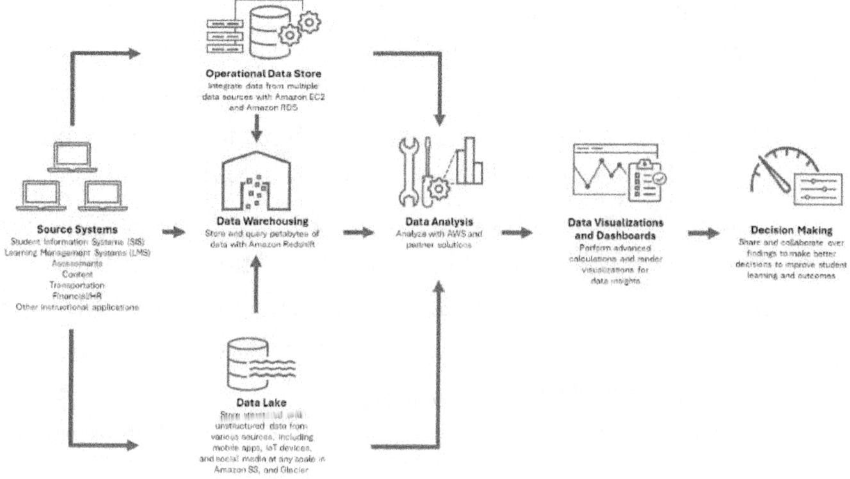

Figure 87 Decision Process

Population Decision Support Systems

The integration of data warehousing and Decision Support Systems (DSS) within the healthcare sector has paved the way for the development of sophisticated Population Decision Support Systems (PDSS). These systems enable the monitoring of performance across various levels of the industry, ultimately enhancing patient care and operational efficiency. After receiving treatment, patients have the opportunity to provide feedback on their experiences through the system, allowing healthcare providers to assess levels of customer satisfaction (George, Kumar & Kumar, 2015).

Patients can share their feedback by either rating their experience on a numerical scale or by sending textual comments that the system can interpret or forward to the customer satisfaction assessment team. This capability ensures that patient voices are heard and considered in ongoing quality improvement efforts. Furthermore, the system accommodates patients' busy schedules by offering appointment slots that are most convenient for them. Through its workflow tools, the PDSS reminds caregivers of patients who need to be contacted for scheduling important appointments, optimizing any tests that might be required during their visits (Roski, Bo-Linn & Andrews, 2014). For example, if test results are anticipated to take time, the system can effectively schedule two separate encounters: one for conducting the tests and another for discussing treatment options based on the results.

For caregivers, the PDSS streamlines workflows, providing clarity on which patients to attend to and at what times. The data warehouse organizes patient visits in a straightforward manner, enabling doctors to manage their caseloads efficiently. At the operational level, management can analyze patient volume data to determine when to increase or decrease staffing levels (George, Kumar & Kumar, 2015). This insight allows healthcare facilities to proactively address staffing needs and intervene effectively when potential issues arise, preventing minor disruptions from escalating into major challenges (McGinley, 2009). Ultimately, a smooth workflow fosters a positive reputation for healthcare centers, enhancing patient trust and satisfaction.

Data Warehouse Viewpoints

Data within a healthcare facility must be integrated safely, accurately, and efficiently, encompassing operational reporting, institution-wide reporting, and outreach opportunities.

Operational Reporting

From the operational reporting perspective, the management team can utilize integrated data systems to identify gaps or problems within healthcare delivery. Once these gaps are pinpointed, management can implement targeted solutions, thereby elevating the quality of care provided by the staff. This viewpoint allows for the comparison of different practices among physicians, facilitating the sharing of best practices among individuals. By identifying and promoting effective approaches, healthcare facilities can abandon less effective procedures, thus driving continuous improvement (Arunachalam, Page & Thorsteinsson, 2017). Additionally, operational reporting enables the evaluation of caregiver performance through DSS, allowing for the recognition of high achievers and providing support for those who may need to improve their performance.

Institution-Wide Reporting

Evaluating the overall performance of a healthcare facility is crucial for benchmarking against other institutions. Institution-wide DSS reports can highlight areas of weakness within the facility, guiding improvements aligned with quality healthcare objectives (Inmon, 2007). Identifying areas requiring additional resources is another critical function of data warehousing, enabling proactive management to avoid negative publicity and ensure the facility's growth. The rapid generation of accurate reports equips management with essential insights when presenting to board members or stakeholders.

Outreach Opportunities

Outreach opportunities facilitate individualized patient care through comprehensive data stored in the data warehouse. DSS provides a framework for alerting both patients and caregivers about important factors such as missed appointments or pending tests. By leveraging personalized data, the system promotes proactive treatment by comparing individual conditions against associated risks and suggesting appropriate preventive measures. This targeted approach not only enhances patient engagement but also fosters better health outcomes by encouraging timely interventions.

In conclusion, Population Decision Support Systems represent a significant advancement in healthcare data management. By leveraging data warehousing and DSS, healthcare providers can enhance patient experiences, improve operational efficiency, and ensure high-quality care. As these systems evolve, they will continue to play a vital role in shaping the future of healthcare delivery, ultimately leading to healthier populations and more effective healthcare systems.

Benefits of Data Warehousing in Healthcare

Quick Access to Desired Data

Data warehousing has revolutionized the way healthcare facilities access and utilize information. With the capability to retrieve vast amounts of data in mere seconds (Inmon, 2007), healthcare managers can generate daily or weekly reports on patient progress. This rapid access allows doctors to be alerted to specific cases requiring immediate attention, fostering a proactive approach to patient care. Additionally, management can monitor compliance with established standards, ensuring that caregivers adhere to protocols designed to enhance patient safety and quality of care.

Establishing and Maintaining a Smooth Workflow

Managing hundreds of patients with diverse needs and schedules can be a daunting task. However, data warehousing supports the establishment of efficient workflows by automating appointment scheduling and reminders. Timely notifications are sent to both patients and healthcare providers, ensuring that encounters are well-prepared and that medications or vaccinations are administered as needed (McGinley, 2009). This automation minimizes the reliance on manual processes, which are often error-prone, thereby enhancing the overall efficiency of healthcare operations. Furthermore, these timely reminders help maintain patient health, particularly in situations where medication shortages may delay treatment.

Identifying Quality of Care Issues

Data warehousing also plays a crucial role in identifying disparities in the quality of care provided to different patient demographics. For example, it can reveal if caregivers spend disproportionate amounts of time with higher-income patients compared to those from lower-income backgrounds (Koh & Tan, 2011). Such inequities are unethical and can be promptly addressed through the insights provided by Data Decision Support Systems (DSS). By fostering equitable care practices, healthcare facilities can ensure that all patients receive the attention and treatment they deserve.

Reduction of Risks Through Early Identification of Risky Areas

The ability to identify **risks early** is another significant advantage of data warehousing. Managers can detect gaps in facility operations and assess potential risks to patient safety (Wang, Kung & Byrd, 2018). For example, discrepancies in the time spent on interpreting incomplete tests during treatment can be highlighted, alerting healthcare providers to potential misdiagnoses that could negatively impact patient outcomes. DSS helps identify these flaws, enabling healthcare facilities to implement corrective measures that enhance patient safety and treatment efficacy.

Increased Performance and Improved Reputation

Effective data warehousing contributes to the overall performance of healthcare facilities, fostering a positive reputation in the community. As performance improves, healthcare facilities often experience an increase in patient volume, leading to better negotiation power when establishing contracts and partnerships. A solid reputation built on reliable data and quality care can attract more patients, further enhancing the facility's standing and sustainability.

Chapter Conclusion

Historically, data warehousing was not integrated into the healthcare sector due to significant differences in data types compared to other industries. Traditional Decision Support Systems were primarily designed for numerical data, rendering them less effective in a healthcare context dominated by textual data. However, advancements in technology have led to the development of systems capable of handling textual data, enabling healthcare facilities to reap the benefits of data integration similar to those seen in other sectors.

Through the integration of DSS, the healthcare industry has embraced initiatives such as proactive treatment programs, chronic disease monitoring, and the establishment of population decision support systems. These initiatives have improved patient communication, enhanced chronic disease management through accurate data collection, and facilitated the monitoring of healthcare provider performance.

In summary, the various perspectives on healthcare data warehousing-including operational, institution-wide, and outreach opportunities-highlight its multifaceted benefits. Quick access to data, efficient workflows, identification of quality care issues, risk reduction, and improved performance are just a few of the advantages. However, challenges remain, particularly regarding data security in the realm of healthcare DSS. As the industry continues to evolve, addressing these challenges will be essential for maximizing the potential of data warehousing in healthcare.

Chapter 18 New Fabric Features

Introduction

Microsoft Fabric regularly introduces new features to expand its capabilities and improve user experiences. The features are usually first shared in preview, allowing users to provide constructive feedback before they become generally available. As of 5th December 2024, more than 50 features had been offered. This report summarizes some of the most important features to help CIO managers make informed implementation decisions.

The Major New Features

Code-First AutoML

This feature is designed to automate machine learning (ML) workflows. It comprises a set of tools and methods that can automatically train and optimize ML models for any data or task. For example, AutoML can choose the right ML model and hyperparameters for different contexts. Users can employ Fabric's MLFlow integration to monitor and analyze their AutoML runs and the flaml.visualization module to draft interactive plots for ML outcomes. As with all premium Fabric features, a subscription is required to access these services.

Code-First Hyperparameter Tuning

FLAML is now integrated into Fabric Data Science to facilitate hyperparameter tuning. The flaml.tune feature is used to streamline this procedure, enabling efficient and cost-effective hyperparameter tuning. Three main steps are included in the workflow: defining tuning objectives, creating a hyperparameter search space, and determining the tuning constraints. The MLFlow integration has also been improved to facilitate more effective monitoring and management of tuning trials. Tuning trials can also be analyzed using the flaml.visualization module. These provisions translate into more insightful and intuitive data exploration and analysis.

Copy Job

Fabric recently introduced Copy Job in Data Factory, which streamlines and improves the user-friendliness of data ingestion. This feature supports different data delivery styles, including incremental and batch copy, simplifying data ingestion for all sources and destinations. Copy job also automatically captures changes in data, minimizing manual oversight and ensuring all available information is current. Users can control how data is moved. For example, read/write behaviors can be configured to suit one's specific needs. Furthermore, Copy Job has a serverless architecture and parallelism at multiple levels for high-performance data movement.

Data Factory Apache Airflow Job

The Data Workflows feature, powered by Apache Airflow, has been introduced in Data Factory to improve the construction and management of code-based data pipelines. This feature is handy for authoring, scheduling, and monitoring functions on Python-based data processes structured as Directed Acyclic Graphs (DAGs). The Data Workflows service provides a software-as-a-service (SaaS)-like experience for working with DAGs in the Apache Airflow environment. It also supports rapid cluster resumption, auto-pause, and autoscaling to improve performance and cost efficiency. Finally, Data Workflows has cloud-based authoring capabilities and is compatible with all Apache Airflow libraries and plugins.

Data Pipeline Capabilities in Copilot for Data Factory

Fabric now includes new Data pipeline capabilities in Copilot for Data Factory, which functions as an AI assistant for managing pipelines. Users can effortlessly create data pipelines through either of two processes: (1) providing one elaborate prompt or (2) engaging with Copilot stepwise to issue detailed prompts at each stage. A "summarize this pipeline" option is included in the starter prompt to derive a clear summary of a targeted pipeline, helping users easily understand their complex data pipelines. Copilot for Data Factory also accurately interprets data pipeline error messages and recommends actionable remedies.

Data Wrangler for Spark DataFrames

Data Wrangler has recently been updated to support pandas and Spark DataFrames. It can now simultaneously produce Python and PySpark code for exploratory data analysis. Spark DataFrames can opened directly from the fabric notebook under the same dropdown prompt that displays pandas DataFrames. While it significantly improves data analysis, Data Wrangler only supports custom code operations for pandas DataFrames.

Data Science AI Skills

AI Skills is a new feature that allows Fabric users to build unique generative AI experiences. It includes multiple provisions designed to optimize workflow and maximize data security. For example, one can specify the data they wish the AI to access in a targeted database. English instructions can then be issued to guide the AI in following definitions and rules. Example questions and query pairs can also be provided to train the AI to answer similar questions. Moreover, AI skills ensures that the Fabric account of the individual asking the question is used to execute all underlying queries.

Dataflow Gen2 with CI/CD and Git integration

Dataflow Gen2 supports Git integration and Continuous Integration/Continuous Deployment (CI/CD). With this feature, users can produce, edit, and manage dataflows in Git repositories that are linked with their Fabric workspaces. The deployment pipeline provision can also be leveraged to automate dataflow deployment between workspaces. Furthermore, the conventional Fabric settings and scheduler can be leveraged to edit and refresh Dataflow Gen2 settings. Users can also directly create their Dataflow Gen2 in a workspace folder.

Delta Column Mapping in the SQL Analytics Endpoint

Delta tables with column mapping have been added to SQL analytics, allowing users to include spaces and characters (,;{}()\n\t=) in column names. These extra characters are captured in lakehouse and conveyed through the SQL Analytics endpoint, the Semantic model, and Power BI reports. This feature activates the SQL Analytics endpoints on schema-enabled lakehouses, facilitating the querying of delta tables in schemas within the SQL analytics endpoint. An improved editor in SQL Analytics endpoint and Fabric Data Warehouse is also included to enhance the experiences of SQL developers.

Domains in OneLake

Fabric includes data mesh architecture to help organize data into domains and filter and discover content by domain. Domains have been introduced as the central enabler for data mesh architecture, availing the requisite infrastructure for a decentralized architecture. This feature also includes optimized consumption for each data business context and federated governance. With these new provisions, data consumers can easily trace content and ensure optimal structuring and control over data.

Eventhouse Query Acceleration for OneLake Shortcuts

Query Acceleration for OneLake Shortcuts has been recently introduced to fast-track ad-hoc queries over data held in OneLake. Eventhouse is fundamentally a database designed to store and manage event-based data. It is suited for data in motion, ensuring optimal indexing and partitioning in storage. Furthermore, Eventhouse is compatible with free text, semi-structured, and structured data. These provisions enable low-latency, high-performance analysis and rapid ingestion and querying. Query acceleration also alters the preexisting structure where queries were implemented over OneLake shortcuts. Specifically, OneLake shortcuts are now structured as references from an Eventhouse that link to the internal Fabric or external resources.

Eventstream Processing and Routing Events to Activator

Eventstream was recently updated to support the use of business requirements to process and transform events. The transformed events are then relayed to the Activator, where the user can define conditions and rules for alerts to monitor events. This feature is fundamentally different from the preexisting Reflex destination for various reasons. For example, Eventstream can transform events before writing them to the destination. Other formats for routing to the destination are also supported, and Activator addresses the "10 events per second" limitation.

High Concurrency Mode for Notebooks in Pipelines

The high concurrency mode has been introduced for pipeline notebooks, providing session-sharing capabilities. With this update, notebooks can automatically be integrated into an active high-concurrency session. This process reduces the session start time to about 5 seconds for shared notebooks, up to 30 times faster than conventional methods. Furthermore, notebooks can rapidly attach pre-warmed high concurrency sessions to an existing Spark session, improving the general pipeline performance. High-concurrency mode also supports session tags, optimizing session management by allowing users to link notebooks with specific high-concurrency sessions. Users are only billed for one session when sharing sessions across different notebooks.

Spark connector for Fabric Data Warehouse in Spark Runtime

Spark connector was recently added to Data Warehouse, allowing data scientists and Spark developers to access and use data from a specific warehouse and a lakehouse's SQL analytics endpoint. This feature offers additional capabilities, such as working with data from a warehouse or lakehouse in one or multiple workspaces. Moreover, the lakehouse's SQL analytics endpoint is automatically discovered for each workspace scenario. Spark connector also retains the SQL engine-level security models, such as column-, row-, and object-level security, when accessing a view or table.

Fabric Apache Spark Diagnostic Emitter

The Apache Spark Diagnostic Emitter allows users to send metrics, event logs, and logs from Spark applications to different destinations, such as Azure Log Analytics, Azure Storage, and Azure Event Hubs. These capabilities improve visibility into the performance of Spark applications by facilitating efficient monitoring and troubleshooting. Users can also easily configure Spark to transmit logs to one or multiple destinations, supporting various related resources, such as Azure Key Vault and connection strings. Numerous metrics and logs can be collected, including detailed Spark application metrics and executor and driver logs.

Fabric SQL Database

An SQL database was recently integrated into Fabric, providing a developer-friendly database for quickly building operational databases. This feature is based on Azure SQL Database, resulting in multiple similarities. For example, the automatic tuning, partitioning, online index operations, database configuration settings, users, and databases from Azure SQL Database are retained. The main security features from Azure SQL Database are also featured, including threat detection, row-level security, dynamic data masking, and application roles.

Folder in Workspace

Workspace cluttering has always been a critical issue in Fabric, severely straining productivity in collaborative settings. The typical workplace gets crowded with hundreds of items as a project progresses. To address this issue, Fabric now includes the Folder feature, an organizational unit that offers a hierarchical structure for arranging and managing all items in a workspace. For example, one can cluster and organize Fabric items based on variables like the deployment stage, analysis category, and data quality.

Iceberg Data in OneLake Using Snowflake and Shortcuts

Microsoft recently partnered with Snowflake, an American cloud-based data storage company, to enable support for Apache Iceberg formatted data in OneLake. The update allows bi-directional data access between Fabric and Snowflake. For example, one can leverage OneLake shortcuts to refer to an Iceberg table drafted using Snowflake. OneLake can also virtualize Iceberg tables as Delta Lake tables for better compatibility with different Fabric engines. Moreover, users can write Iceberg tables directly to OneLake.

Real-Time Intelligence

Real-Time Intelligence helps organizations build end-to-end solutions harnessing the power of real-time granular data at scale. This feature allows businesses to easily ingest, process, analyze, and visualize data and monitor and act on events. For example, manufacturing companies can exploit real-time signals to improve supply chain efficiency and production quality. Similarly, logistics firms can enhance delivery, routing, and staffing. The main components in Real-Time Intelligence that enable these uses include the

Activator, Real-Time Dashboards, Eventhouse, Improved Eventstream, and Real-Time Hub.

Workspace Monitoring

Workspace monitoring is a database in Fabric that amasses and organizes metrics and logs from different items in a workspace, enabling easy access and analysis. The feature also creates an Eventhouse database in a user's workspace to automatically collect and organize metrics and logs, facilitating optimal monitoring. Notably, the database takes a read-only structure and is only accessible to users with at least a contributor role in the workspace. The database is also directly accessible from the workplace, and users can develop and save query sets and dashboards for optimal data exploration.

OneLake External Data Sharing

The external data-sharing feature allows users to share data between different Fabric tenants. Data is essentially shared in-place from OneLake repositories in the sharer's Fabric tenant, meaning it is not copied to the recipient's tenant. Instead, the sharing feature creates a OneLake shortcut in the recipient's tenant that references the original data. Furthermore, all externally shared data is read-only and can be consumed by all OneLake-compatible workloads.

Despite the underlying similarities, this external data-sharing feature is unrelated to the previously existing processes for sending Power BI semantic models to Microsoft Entra B2B guest users.

Fabric Workload Development Kit

Fabric also recently introduced the Workload Dev Kit to expand workloads and provide a robust toolkit for designing, developing, and interoperating with Fabric using backend RESTful Application Programming Interfaces and frontend Software Development Kits. This feature allows independent software vendors and developers to natively integrate existing and new applications into the Fabric workload hub. Developed solutions can also be shared through the Azure marketplace.

Notebook Git Integration

Fabric introduced notebook Git integration and deployment pipeline, facilitating seamless incorporation and implementation of notebooks with Git. The notebook Git integration allows the attached environment's mapping relationship to persist when syncing to a new workplace. Thus, if the notebook and environment are jointly committed to the Git repository and synced to a different workplace, the new environment and notebook will be bound. Similarly, the deployment pipeline binds the default lakehouse and attached environment in a common workspace when deploying to a different stage.

New Features Summary

This report provides a detailed overview of the latest Fabric features, tailored to help management professionals, particularly CIOs, make informed implementation decisions. The presented features primarily involve advancements in machine learning, robust data pipeline management tools, integrations for real-time intelligence, and collaborative enhancements, such as external data sharing and Git compatibility. Furthermore, the features collectively address critical needs like optimizing resource usage, enabling real-time insights, ensuring data governance, and simplifying complex operations. However, it is worth noting that most features are still in preview mode and should transition to general availability in the coming months. By leveraging these innovations, organizations can accelerate digital transformation, improve productivity, and maintain competitive agility in handling evolving data and AI-driven processes.

Conclusion

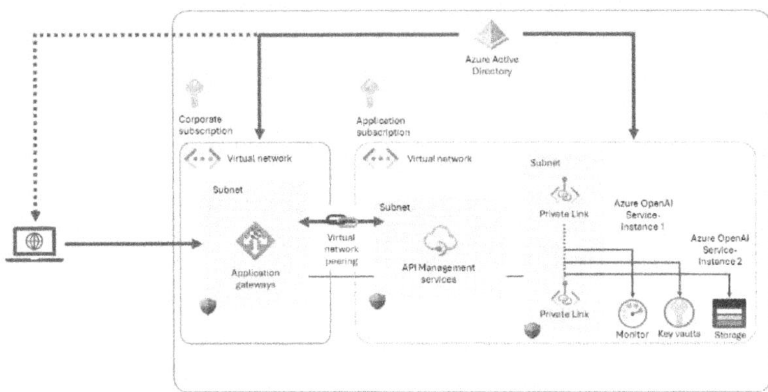

Figure 88 Fabric Architecture

Microsoft Fabric is a transformative product designed to help businesses effectively manage and understand Big Data in today's complex landscape. It offers a comprehensive and seamless analytics platform equipped with a full suite of tools tailored to meet the diverse needs of organizations, regardless of their size or industry.

One of the standout features of Fabric is its ability to integrate various analytics services, which significantly reduces the complexities and costs typically associated with data analytics. This integration not only streamlines processes but also makes powerful analytics more affordable and accessible for enterprises and individual users alike.

Furthermore, Fabric's compatibility with Microsoft 365 enhances its general availability and usability, allowing users to leverage familiar tools within a unified environment. This synergy empowers organizations to collaborate more effectively, driving better insights and decision-making.

Additionally, the incorporation of artificial intelligence into Fabric's framework enables customers to extract maximum value from their data. AI-driven insights assist businesses in identifying trends, making predictions, and optimizing operations, ultimately leading to more informed strategic decisions.

Overall, these features position Microsoft Fabric as a leading solution in data management and analytics. Its commitment to simplifying the complexities of Big Data while enhancing accessibility and usability makes it an invaluable asset for organizations striving to harness the power of their data in the modern business landscape. As the demand for effective data solutions continues to grow, Microsoft Fabric stands out as a pivotal tool for navigating the challenges and opportunities of data-driven decision-making.

References

Alzubaidi, L., Zhang, J., Humaidi, A. J., Al-Dujaili, A. Q., Duan, Y., Al-Shamma, O., Santamaría, J., Fadhel, M. A., Al-Amidie, M., & Farhan, L. (2021). Review of deep learning: concepts, CNN architectures, challenges, applications, future directions. Journal of Big Data, 8(1). https://doi.org/10.1186/s40537-021-00444-8

Alzubaidi, L., Zhang, J., Humaidi, A. J., Al-Dujaili, A. Q., Duan, Y., Al-Shamma, O., Santamaría, J., Fadhel, M. A., Al-Amidie, M., & Farhan, L. (2021). Review of deep learning: concepts, CNN architectures, challenges, applications, future directions. Journal of Big Data, 8(1). https://doi.org/10.1186/s40537-021-00444-8

Cáliz, D., Samaniego, G., & Cáliz, R. (2016). Methodological Proposal of Policies and Procedures for Quality Assurance in Information Systems for Software Development Companies Based on CMMI. JSW, 11(3), 230-241.

Chandrasekara, C., & Herath, P. (2019). Hands-on Azure Repos: Understanding Centralized and Distributed Version Control in Azure DevOps Services. Apress.

Gundu, S. R., Panem, C. A., & Thimmapuram, A. (2020). The Dynamic Computational Model and the New Era of Cloud Computation Using Microsoft Azure. SN Computer Science, 1(5), 264. https://doi.org/10.1007/s42979-020-00276-y

Gustafsson, N. (2023, May 23). Introducing Synapse Data Science in Microsoft Fabric. Microsoft. https://blog.fabric.microsoft.com/en-us/blog/introducing-synapse-data-science-in-microsoft-fabric/

Heller, P. (2021). Automating Workflows with GitHub Actions: Automate software development workflows and seamlessly deploy your applications using GitHub Actions. Packt Publishing Ltd.

Huawei Technologies Co., Ltd. (2022). A General Introduction to Artificial Intelligence. In Artificial Intelligence Technology (pp. 1-41). Singapore: Springer Nature Singapore. https://doi.org/10.1007/978-981-19-2879-6_1

Huawei Technologies Co., Ltd. (2022). A General Introduction to Artificial Intelligence. In Artificial Intelligence Technology (pp. 1-41). Singapore: Springer Nature Singapore. https://doi.org/10.1007/978-981-19-2879-6_1

Hundhausen, R. (2021). Professional Scrum Development with Azure DevOps. Microsoft Press.

IBM. (2023). AI vs. Machine Learning vs. Deep Learning vs. Neural Networks: What's the difference? IBM Blog. https://www.ibm.com/blog/ai-vs-machine-learning-vs-deep-learning-vs-neural-networks/

IBM. (2023). AI vs. Machine Learning vs. Deep Learning vs. Neural Networks: What's the difference? IBM Blog. https://www.ibm.com/blog/ai-vs-machine-learning-vs-deep-learning-vs-neural-networks/

Jagdale, K. R., Shelke, C. J., Achary, R., Wankhede, D. S., & Bhandare, T. V. (2022). Artificial intelligence and its subsets: machine learning and its algorithms, deep learning, and their future trends. Int. J. Emerg. Technol. Innov. Res, 9(5).

Jagdale, K. R., Shelke, C. J., Achary, R., Wankhede, D. S., & Bhandare, T. V. (2022). Artificial intelligence and its subsets: machine learning and its algorithms, deep learning, and their future trends. Int. J. Emerg. Technol. Innov. Res, 9(5).

Kromer, M. et al. (2023, November 15). What is Data Factory in Microsoft Fabric? Microsoft. https://learn.microsoft.com/en-us/fabric/data-factory/data-factory-overview

Lucznik, J. (2023, May 23). Introducing Synapse Data Engineering in Microsoft Fabric. Microsoft. https://blog.fabric.microsoft.com/en-us/blog/introducing-synapse-data-engineering-in-microsoft-fabric/

Makhija, J. (2023, August 11). Navigating the data platform landscape: Finding the perfect fit for your company Azure Synapse or Microsoft Fabric. https://community.dynamics.com/blogs/post/?postid=4c923e38-8738-ee11-bdf4-000d3a4e511f

Manis, K. (2023, May 23). Introducing Microsoft Fabric and Copilot in Microsoft Power BI. Microsoft. https://powerbi.microsoft.com/en-us/blog/introducing-microsoft-fabric-and-copilot-in-microsoft-power-bi/

Microsoft Build. (2024). What is Microsoft Fabric? https://learn.microsoft.com/en-us/Fabric/get-started/microsoft-Fabric-overview

Microsoft. (n.d.). Azure Service Fabric documentation. https://learn.microsoft.com/en-us/azure/service-fabric/

Mohammed (2023). Unveiling Microsoft Fabric: Data Analytics for the Era of AI. Retrieved from https://medium.com/@techlatest.net/unveiling-microsoft-Fabric-data-analytics-for-the-era-of-ai-d7a8f1be4be0

Mowad, A. M., Fawareh, H., & Hassan, M. A. (2022, November). Effect of using continuous integration (ci) and continuous delivery (cd) deployment in DevOps to reduce the gap between developer and operation. In 2022 International Arab Conference on Information Technology (ACIT) (pp. 1-8). IEEE. https://doi.org/10.1109/ACIT57182.2022.9994139

Opara-Martins, J. (2023). Perspective Chapter: Cloud Lock-in Parameters–Service Adoption and Migration. In Edge Computing-Technology, Management, and Integration. IntechOpen. DOI: 10.5772/intechopen.109601

Peltier, T. R. (2016). Information Security Policies, Procedures, and Standards: guidelines for effective information security management. Auerbach Publications.

Press, G. (2023). 12 AI milestones: 4. MYCIN, an expert system for infectious disease therapy. Forbes. https://www.forbes.com/sites/gilpress/2020/04/27/12-ai-milestones-4-mycin-an-expert-system-for-infectious-disease-therapy/?sh=32c5ca9076e5

Press, G. (2023). 12 AI milestones: 4. MYCIN, an expert system for infectious disease therapy. Forbes. https://www.forbes.com/sites/gilpress/2020/04/27/12-ai-milestones-4-mycin-an-expert-system-for-infectious-disease-therapy/?sh=32c5ca9076e5

Pugliese, R., Regondi, S., & Marini, R. (2021). Machine learning-based approach: global trends, research directions, and regulatory standpoints. Data Science and Management, 4, 19–29. https://doi.org/10.1016/j.dsm.2021.12.002

Pugliese, R., Regondi, S., & Marini, R. (2021). Machine learning-based approach: global trends, research directions, and regulatory standpoints. Data Science and Management, 4, 19–29. https://doi.org/10.1016/j.dsm.2021.12.002

PWC (2024). 5 ways Microsoft Fabric can help get your data ready for GenAI. Retrieved https://www.pwc.com/us/en/tech-effect/ai-analytics/gen-ai-with-microsoft-Fabric.html

Radanliev, P. (2024). Artificial intelligence: reflecting on the past and looking towards the next paradigm shift. Journal of Experimental and Theoretical Artificial Intelligence, 1–18. https://doi.org/10.1080/0952813x.2024.2323042

Radanliev, P. (2024). Artificial intelligence: reflecting on the past and looking towards the next paradigm shift. Journal of Experimental and Theoretical Artificial Intelligence, 1–18. https://doi.org/10.1080/0952813x.2024.2323042

Safa, N. S., Von Solms, R., & Furnell, S. (2016). Information security policy compliance model in organizations. Computers & security, 56, 70-82.

Sahay, R. (2020). Microsoft Azure architect technologies study companion: Hands-on preparation and practice for exam AZ-300 and AZ-303. Apress.

Sarker, I. H. (2021a). Machine learning: algorithms, Real-World applications and research directions. SN Computer Science, 2(3). https://doi.org/10.1007/s42979-021-00592-x

Sarker, I. H. (2021a). Machine learning: algorithms, Real-World applications and research directions. SN Computer Science, 2(3). https://doi.org/10.1007/s42979-021-00592-x

Sarker, I. H. (2021b). Deep Learning: a comprehensive overview on techniques, taxonomy, applications and research directions. SN Computer Science, 2(6). https://doi.org/10.1007/s42979-021-00815-1

Sarker, I. H. (2021b). Deep Learning: a comprehensive overview on techniques, taxonomy, applications and research directions. SN Computer Science, 2(6). https://doi.org/10.1007/s42979-021-00815-1

Saroar, S. G., & Nayebi, M. (2023, June). Developers' perception of GitHub Actions: A survey analysis. In Proceedings of the 27th International Conference on Evaluation and Assessment in Software Engineering (pp. 121-130). https://doi.org/10.1145/3593434.3593475

Sathy, P. (2023, May 23). Introducing Synapse Data Warehouse in Microsoft Fabric. Microsoft. https://blog.fabric.microsoft.com/en-US/blog/introducing-synapse-data-warehouse-in-microsoft-fabric/

Schuster, Y. et al. (2023, December 14). What is Real-Time Analytics in Fabric? Microsoft. https://learn.microsoft.com/en-us/fabric/real-time-analytics/overview

Seminger, D. et al. (2023, November 16). What is Data Activator? Microsoft. https://learn.microsoft.com/en-us/fabric/data-activator/data-activator-introduction

Senft, S., Gallegos, F., & Davis, A. (2016). Information technology control and audit. Auerbach publications.

Seremet, Z., & Rakic, K. (2021). Best approach to security in Azure DevOps. DAAAM International Scientific Book.

Shahin, M., Babar, M. A., & Zhu, L. (2017). Continuous integration, delivery, and deployment: A systematic review on approaches, tools, challenges, and practices. IEEE Access, 5, 3909-3943. http://dx.doi.org/10.1109/ACCESS.2017.2685629

Shao, F., & Shen, Z. J. (2023). How can artificial neural networks approximate the brain? Frontiers in Psychology, 13. https://doi.org/10.3389/fpsyg.2022.970214

Shao, F., & Shen, Z. J. (2023). How can artificial neural networks approximate the brain? Frontiers in Psychology, 13. https://doi.org/10.3389/fpsyg.2022.970214

Singh, A., & Mansotra, V. (2021). A Comparison on Continuous Integration and Continuous Deployment (CI/CD) on Cloud Based on Various Deployment and Testing Strategies. International Journal for Research in Applied Science and Engineering Technology, 9(VI), 4968-4977.

Solid Soft Reply. (n.d.). Developing applications using microservices and Microsoft Azure Service Fabric. https://www.reply.com/solidsoft-reply/en/Shared%20Documents/Microservices-and-Microsoft-Azure-Service-Fabric-Solidsoft-Reply.pdf

Soni, M. (2015, November). End-to-end automation on cloud with build pipeline: the case for DevOps in insurance industry, continuous integration, continuous testing, and continuous delivery. In 2015 IEEE International Conference on Cloud

Computing in Emerging Markets (CCEM) (pp. 85-89). IEEE. https://doi.org/10.1109/CCEM.2015.29

Taye, M. M. (2023). Understanding of Machine Learning with Deep Learning: Architectures, Workflow, Applications and Future Directions. Computers, 12(5), 91. https://doi.org/10.3390/computers12050091

Taye, M. M. (2023). Understanding of Machine Learning with Deep Learning: Architectures, Workflow, Applications and Future Directions. Computers, 12(5), 91. https://doi.org/10.3390/computers12050091

Tirgari, V. (2012). Information technology policies and procedures against unstructured data: A phenomenological study of information technology professionals. Journal of Management Information and Decision Sciences, 15(2), 87.

Ulagaratchagan, A. (2023, May 23). Introducing Microsoft Fabric: Data analytics for the era of AI. Microsoft. https://azure.microsoft.com/en-us/blog/introducing-microsoft-fabric-data-analytics-for-the-era-of-ai/

Vasilescu, B., Van Schuylenburg, S., Wulms, J., Serebrenik, A., & van den Brand, M. G. (2014, September). Continuous integration in a social-coding world: Empirical evidence from GitHub. In 2014 IEEE International Conference on Software Maintenance and Evolution (pp. 401-405). IEEE. https://doi.org/10.1109/ICSME.2014.62

Vishwakarma, S. (2023). Introduction to GitHub Actions: Streamlining CI/CD Pipelines. Documatic. https://dev.to/documatic/introduction-to-github-actions-streamlining-cicd-pipelines-1fkf

Weber, I., Nepal, S., & Zhu, L. (2016). Developing dependable and secure cloud applications. IEEE Internet Computing, 20(3), 74-79. http://dx.doi.org/10.1109/MIC.2016.67

Zampetti, F., Bavota, G., Canfora, G., & Di Penta, M. (2019, February). A study on the interplay between pull request review and continuous integration builds. In 2019 IEEE 26th International Conference on Software Analysis, Evolution, and Reengineering (SANER) (pp. 38-48). IEEE. https://doi.org/10.1109/SANER.2019.8667996

Zeeshan, A. A., & Zeeshan, A. A. (2020). Securing Build Systems for DevOps. DevSecOps for. NET Core: Securing Modern

Software Applications, 163-214. https://doi.org/10.1007/978-1-4842-5850-7_5

www.ingramcontent.com/pod-product-compliance
Lightning Source LLC
Chambersburg PA
CBHW051411050726
47595CB00010B/4015